Some Mathematical Questions in Biology. VII

Lectures on Mathematics in the Life Sciences
Volume 8

Some Mathematical Questions in Biology. VII

The American Mathematical Society
Providence, Rhode Island
1976

Proceedings of the Ninth Symposium on
Mathematical Biology held in New York, January, 1975.

edited by
Simon A. Levin

Library of Congress Catalog Card Number 77-25086
International Standard Book Number 0-8218-1158-4
AMS 1970 Subject Classification: 92A05

The Symposium was sponsored by the
National Science Foundation under Grant No. MPS-75-05411

CONTENTS

CONTENTS

FOREWORD

This volume contains lectures given at the Ninth Symposium on Some Mathematical Questions in Biology, held in New York on January 29-30, 1975, in con-
junction with the annual meeting of the American Association for the Advancement
of Science. The Symposium was co-sponsored by the American Mathematical Society
and by the Society for Industrial and Applied Mathematics under the auspices of
Section A, Mathematics, of the AAAS.

The first three papers in this volume, by Simon Levin, George Oster, and
Brian Charlesworth, deal with problems in ecology and evolutionary biology.
The classical mathematical theory of ecological dynamics has for the most part
treated the population as a collection of identical individuals, each with the
same requirements and potential for growth. However, real populations of this
kind are hard to find; individuals differ, with regard to age, size, spatial
location, social rank, and other traits. The first three papers relax in
various ways the constrictive assumption of homogenity, and place emphasis on
the critical role of population structure. The paper by Levin introduces spatial structure, and investigates the changes which occur in ecological theory.
In the second paper, Oster surveys the implications of the introduction of age
structure; and in the third, using a somewhat different framework, Charlesworth
explores the interface between demographic and evolutionary parameters.

The second half of the volume deals with problems in neurobiology. Cohen
and Rinzel survey a large body of the literature concerning mathematical contributions to neurophysiology, beginning from the cellular level and classical
Hodgkin-Huxley theory for the transmission of a voltage pulse and proceeding to
more modern theory and work on dendritic integration. Barlow, in the final
paper, provides a counterpoint with a somewhat more negative view of the role
of mathematics and a caveat that to make an impact the mathematician must

become a neurobiologist.

It is a pleasure to acknowledge the support provided by the National
Science Foundation.

SIMON A. LEVIN
CORNELL UNIVERSITY
September, 1975

Lectures on Mathematics in the Life Sciences
Volume 8, 1976

SPATIAL PATTERNING AND THE STRUCTURE OF ECOLOGICAL COMMUNITIES

Simon A. Levin

1. Introduction

Fifty years ago, mathematical ecology had its genesis through the coincident efforts of two intellectual giants, Alfred Lotka and Vito Volterra. Their works were the first major attempts to represent and analyze ecological interactions within the framework of dynamical systems theory, and were so imposing that later generations have treated them as revelation rather than commentary. The unfortunate consequence has been that subsequent mathematical enquiries have been heavily influenced towards the detailed analysis of a class of equations which should have served as metaphors, and progress in new directions has been restricted.

The contributions of Lotka and Volterra began in a vacuum, virtually created a new field, and have obviously withstood the test of time. Their work emphasized the importance of models, which can serve not only as tools for prediction but also to stimulate new avenues for experimentation. However to relate to the research of the experimental ecologist, they must be expanded beyond the classical forms, and I shall attempt to identify some of the directions in which expansion must proceed. In particular, I shall focus on the kinds of general phenomena which arise when spatial structure is introduced, concentrating on "robust" results; i.e., those which are not tightly tied to very specific details of particular models.

The first three papers in this volume relax in various ways one of the fundamental constraints of the classical literature, the assumption of population homogeneity with respect to all characteristics. Population structure, by which I mean the distribution within a population of such properties as

genotypic and phenotypic traits, age, spatial location, and social role, is
not just the fine tuning of population theory. The description and elaboration of the ecological and adaptive significance of the various aspects of
population structure are the fundamental questions of population biology. Of
particular importance is an understanding of the interaction between structure and dynamics. Further, it is essential to determine to what degree the
elaborate ecological theory which has been founded upon the Lotka-Volterra
approach is altered when various aspects of population structure are introduced. In particular, structural feedback loops may increase the diversity
of limiting factors in a community (Levin 1970) and thereby permit survival
of populations which would otherwise be competitively excluded. This idea
has been developed to some extent with respect to genetic structure (Pimentel
1961, 1968; Levin 1972, 1973; León 1974; Pimentel, Levin, and Soans 1975;
Levin and Udovic 1975; Udovic and Levin 1975), but applies as well with regard to space (Levin 1974a)and age (Ayala 1971,Demetrius 1975,Skellam 1951).

The focus of this paper is the role of spatial structure; and I shall
avoid explicit discussion of age structure and genetics, topics which are
covered elsewhere in this volume (Charlesworth 1975, Oster 1975) and in Levin
and Udovic (1975) and Udovic and Levin (1975). However, since dispersal is
often confined to particular stages of organism life cycles, age structural
considerations lie in the background of the discussion of spatial structure.
Further, although for most purposes it is convenient to assume that the evolutionary time scale is long compared to the ecological, a comprehensive theory
would include all structural aspects. Species evolutionarily relate to the
heterogeneity or local unpredictability of environment by dispersal to
"spread the risks" (den Boer 1968, 1971, Reddingius 1971, Reddingius and den
Boer 1970, Roff 1974, Smith 1972, Strathmann1974), and the adaptive significance of such strategies is of fundamental importance to evolutionary theory
(Strathmann1974, Gadgil 1971). Thus, beyond the considerations of this paper,
one must turn attention to the relations between spatial structure, age structure, and evolution.

2. Preliminary considerations

Space complicates ecological interactions by two fundamental mechanisms: it allows for non-uniform patterns of environment and population density, and for movement (dispersal or migration) of individuals or their gametes from one location to another. The classical modelling approach focuses on populations at a point in space, ignoring movement; or equivalently (from the point of view of the mathematics) it assumes perfect mixing over a region. By restricting the species' ecological and evolutionary arena in this way, one ignores essential aspects of the species' response to the full range of environments it encounters (its ecotope, sensu Whittaker, Levin, and Root 1973, 1975). As such, it describes a situation more appropriate to the laboratory than to the natural environment. Moreover, even in the laboratory, spatial heterogeneity may be essential for continued coexistence of populations (Huffaker 1958, Pimentel 1965, Luckinbill 1974, Smith 1972).

Space serves in a variety of ways to facilitate coexistence of competitors, of predator and prey, and of parasite and host, and in general to increase diversity. I seek in this paper to catalogue some of these ways, and shall classify them (somewhat arbitrarily) as due to: local uniqueness, phase difference, or dispersal. These categories overlap, but the distinctions generally are clear and this gross taxonomy provides a convenient starting point.

By local uniqueness, I refer to the distinctive destiny (within the context of a particular time scale) that would lie in store for each local environment were it to be suddenly sealed off from the rest of the universe. Local uniqueness is defined by a combination of habitat variation and chance events, including colonization episodes. On a longer time scale, the same locale may be exposed to different conditions and so the local uniqueness of an environment is temporal as well as spatial.

Variation in system description (e.g., variation in system parameters) over space and time is the most obvious component of local uniqueness. In the classical viewpoint, microhabitat variation in parameters gives rise to

the notion of the polyclimax (Whittaker 1974) as a mosaic of equilibrium
states each related to local climatic (and edaphic) parameters; macrohabitat
variation explains the differences between the climaxes achieved in particular habitats.

Above and beyond such variation, however, are the effects due to chance
events. The vagaries of colonization episodes, for example, determine the
individual trajectory which describes the transient aspects of the dynamics
following a disturbance (e.g., secondary succession pattern Horn 1974). Further,
the possible existence of multiple stable attractors in the dynamics (Park
1962, Lewontin 1969, Sutherland 1974) means that random events may influence
qualitatively the asymptotic behavior of the system.

All of these contribute to what I identify as local uniqueness.

Transient events are of secondary importance in the classical climax
view (Clements 1928) of vegetation, or in the equilibrium view of competition
which pervades mathematical theory; but they assume new importance when one
recognizes the climax as a "climax-pattern" which includes various successional states (Whittaker 1974, Horn 1975, 1976). Recurrent disturbance is
probably an important component of most systems (Heinselman 1973, Levin and
Paine 1974, 1975, Commito and Rubenstein 1975, Loucks 1970, Wright 1974), and
prevents a system from ever attaining a monotonous homogeneous steady state.
When disturbance is localized and asynchronized regionally, additional micro-
variation enters the pattern due to phase difference or maturity, essentially
time since last disturbance. Even microhabitats which are destined (in the
sense of local uniqueness) to share the same pattern of development may still
differ at any point in time due to difference in timing of disturbance. The
effect of this is to maintain a higher system diversity at any point in time
by assuring that more successional states are represented (as in the climax-
pattern) (Whittaker 1974, Horn 1975, 1976, Levin and Paine 1974, 1975, Loucks
1970).

Finally, dispersal per se may play an important role in preserving cer-

tain species in the system. Most obviously, some species are maintained only
by dispersal into the system from source areas by migrants, pollen, seeds, or
larvae. However, even within a closed system, dispersal may provide the key
to survival either through spatio-temporal "fugitive" strategies or by time-
independent spatial ones. The former is more familiar, with fugitives being
maintained globally despite ecological interactions which doom them locally
(Huffaker 1958, Dodd 1940, 1959, Hutchinson 1959). The ability of the population to "disperse" is the key to survival. Indeed, dispersal and the difference in timing (phase difference) of local habitats provide the ingredients
which permit a species to achieve a balance between dispersal ability and competitive ability unique to its own evolutionary role, and is central to the
survival of _most_ species. It is perhaps unconventional to view the fugitive
as the rule rather than the exception, but it is the case that most species
are to some degree locally ephemeral.

Finally, the importance of differential dispersal abilities of species in
establishing time-independent non-uniform spatial patterns has been suggested
as a possible explanation of patchiness, for example in the plankton(Okubo
1974, Segel and Levin 1975, Levin and Segel 1975); and Steele (1973,1974a,b)
has put forward a rather different hypothesis, also dependent on the inter-
play of dispersal and biological interactions.

3. Mathematical framework

In this section, a general mathematical structure is developed in which
the concepts discussed in the preceding section may be formalized, and this
framework provides the beginnings of a quantitative theory. It is, however,
only a point of departure; for specific applications, it may need to be either
specialized or extended.

Whatever formalism one uses should account for three basic kinds of dynamics, those due to:

(1) Proximal interactions - that is, interactions between organisms
within a particular local region. Essentially, these are modelled by local-

ized growth equations which it is assumed can be described entirely in terms
of local parameters. "Proximal" is employed here entirely in the spatial context, so that non-proximal temporal effects may be included (through parameter
variation) in the local description. So indeed may localized descriptors of
population structure, including age or genetic. Thus the equations in question may be of a variety of types - for example differential, difference, dif-
ferential-difference, integro-differential, or partial differential - and may
include stochastic elements. For definiteness, the skeletal framework which
I shall develop uses differential equations, but this is only a detail of the
fine structure of the equation system.

"Local" is clearly an ambiguous term. In one formulation, hereafter referred to as discrete, the environment is viewed as a mosaic of patches or
cells within each of which perfect mixing is assumed. In this usage, "local"
refers to interactions within a single patch, and the local dynamics are those
which would result if organisms were confined to that patch. Patches need be
neither contiguous nor identical in size.

An alternative to the discrete is the continuous description, in which
the environment is viewed as a continuum. The continuous model can be obtained as the limit of a sequence of discrete models in each of which all
patches are of uniform size and the state variables are intensively defined
(e.g., as concentrations or densities). Under appropriate assumptions, as the
standard patch size becomes arbitrarily small the discrete local dynamics tend
to a limiting form, which provides the local description in the continuous
case.

(2) Movement (exchange) between locales - Again, the specifics can take
several forms, with the null hypothesis being passive diffusion; convective
and gradient-oriented movement are, however, likely to be of importance. Age-
dependent or genotypic-dependent dispersal is best dealt with by the inclusion
of appropriate structure in the localized equations.

(3) Exchange with the outside world - For example, for many systems,

invasions and colonizations, nutrient runoffs, etc., from outside the system provide a major organizing force. By choosing one's reference system of interest, one automatically defines the "outside world" or "bath"as its complement.

As a starting point for the full model, I choose the discrete spatial formulation and continuous time, and wish to write equations for the dynamics of n species distributed over an interconnected network of m patches. Related treatments of population processes in patchy environments may be found in Karlin and McGregor (1972), Chewning (1975), Allen (1975), Maynard Smith (1974), van der Meer (1973), Cohen (1970), Levins and Culver (1971), Horn and MacArthur (1972) and Slatkin (1974). Following Othmer and Scriven (1971, 1974) and Levin (1974a) but with slight changes, the density of species (biological or otherwise) i $(i = 1,\ldots,n)$ in patch μ $(\mu = 1,\ldots,m)$ is denoted u_i^μ. The general model has the form of a system of differential equations

$\quad du_i^\mu/dt$ $=$ {localized growth rate, a function of local densities
$\qquad\qquad\qquad$ and parameters}

$\qquad\qquad$ + {net exchange with other patches}$\qquad\qquad\qquad$(1)

$\qquad\qquad$ + {net exchange with "bath"} ;

that is,

$\quad du_i^\mu/dt$ $=$ $f_i^\mu(\underset{\sim}{u}^\mu,\underset{\sim}{v}^\mu)$ + {net exchange with other patches}

$\qquad\qquad\qquad$ + {net exchange with bath} ,

in which $\underset{\sim}{u}^\mu$ is the vector $(u_1^\mu,\ldots,u_n^\mu)$ of densities of all species or other components (e.g., resources) in patch μ and $\underset{\sim}{v}^\mu$ is a vector of environmental parameters (usually climatic or edaphic, but possibly including species densities). The functions f_i^μ are arbitrary, except that they are assumed to be defined and continuously differentiable on an open set the projection of which onto "$\underset{\sim}{u}^\mu$-space" contains the set $R_\mu:\{u_i^\mu \geq 0 \,\forall\, i\}$; and further, for $\underset{\sim}{u}^\mu\epsilon R_\mu$, $f_i^\mu(\underset{\sim}{u}^\mu,\underset{\sim}{v}^\mu) \geq 0$ if $u_i^\mu = 0$. The definition of R_μ includes all biologically meaningful values of species densities, and the second restriction simply states that non-existent species do not decline in density.

A more realistic description of local dynamics might replace the system
above by a system of stochastic differential equations; but again, the point
of view will be to regard that as an extension of the skeletal model. None-
theless, some aspects of stochastic variation may still appear through the
parametric dependence and initial conditions.

A few words are in order concerning parametric variation. I assume,
first of all, that parameter values are not significantly affected by the
variables u_i^μ. Those parameters for which this is not true are included explicitly as u-variables, and hence this latter set may include things other
than species densities. (In some cases this may mandate modifications in the
assumptions made above.) On the other hand, parametric variation may affect
local dynamics in a variety of ways and on a variety of time scales. The importance of a particular parameter involves both the rate at which it varies
and the degree to which local rates of change of the u-variables depend upon
it.

For parameters whose effect on system dynamics is on a much longer time
scale than interspecific interactions, results such as those of Tikhonov (1948,
1952)(see also Göbber and Seelig 1975) indicate when it is appropriate to view the
parameters as constant, determine the asymptotic behavior pattern, and then
examine the changes in that pattern as the parameters are allowed to vary.
This may be expected to result in only slight variation in the system phase
portrait except for critical "catastrophic" parameter values, where the system
description is "structurally unstable" (Thom 1970). The usual presentation of
this is that parameters change gradually and reach threshold values where instability sets in; but the effect is little different if certain key parameters change very slowly most of the time, but very rapidly for very short periods of time. In many systems, individual patches will reach an ecological
equilibrium or else a plateau which changes very little until intrinsic or
extrinsic disturbances cause total or partial extinction of the biota. Such
an event may in most cases be viewed as due to catastrophic parametric varia-

tion of one kind or another.

For parameters which affect dynamics on a more rapid time scale, the result is that the system loses its autonomy - that is, it must be modelled by time-varying differential equations rather than autonomous ones. The situation becomes totally "non-equilibrium" (Hutchinson 1959).

Finally, if there are parameters which affect system dynamics on a more rapid time scale than the dynamic interactions themselves, the system dyanmics will usually respond only to statistical properties of the parametric variation, and these become the essential parameters of interest.

This completes the description of the first term on the right of (1). By having restricted the problem of exchange with other systems to net exchange with a bath, I have really eliminated the need for a separate bath term;inputs and outputs can be included as parameters in the localized growth rates, although in the sequel we shall often find it convenient to identify the input from the bath explicitly. Also, in more general contexts, it may not be appropriate to view the external world as a bath; i.e., as unaffected by what flows to it from the principal system of interest. In such cases, one must allow for feedback relationships. For example, in considering colonization phenomena, determination of whether it is appropriate to ignore the contribution of the particular area in question to the pool of colonizers is a question of scale; that is, of the fraction of the pool which actually comes from the system under study.

I come now to the difficult question of the exchange rates between patches, difficult largely because of the paucity of quantitative information available about dispersal in natural populations. The simplest assumption possible is the null hypothesis of random movement and passive diffusive-type dispersal. This results in a mathematical expression for movement of u_i from patch μ to patch ν which is simply proportional to u_i^μ. The constant of proportionality will in general vary as a function of both μ and ν. The second term on the right of (1) would then become a sum of terms of the form

$d_{\mu\nu}u_i^\nu$ representing inputs to μ, minus a sum of terms $d_{\nu\mu}u_i^\mu$ representing fluxes from μ to other patches. The coefficients correct for the relative proximities of patches. If movement is truly random, one expects that the d values are symmetric $(d_{\mu\nu}=d_{\nu\mu})$, so that there is no net movement between patches which are at the same density.

As stated earlier, the assumption of passive dispersal is probably little more than a null hypothesis, except in some special situations. Movement of individuals and gametes may be highly dependent upon wind, water, and animal vectors. Moreover, even organisms as tiny as planktonic larvae, cellular acrasioles, and protozoa orient towards various kinds of gradients, including chemical agents and temperature (Adler 1966; Chet, Henis, and Mitchell 1973, Keller and Segel 1970.) Pheromones and other chemicals serve as signals to the existence of conspecifics or potential prey; and whereas the chemicals themselves may spread passively (for the continuous case, see the discussion in Rubinow 1973), organisms following chemical gradients do not. A general approach to the problem of modelling chemotaxis in the continuous context may be found in Keller and Segel (1970). Further, organisms certainly orient towards other cues, including visual ones, and so for example "prey-taxis" may become an essential consideration.

The associated continuous theory is at least as complicated, and some discussion may be found in Skellam (1951), Levin (1974a), Segel and Jackson (1972), Lin and Segel (1974), and Segel and Levin (1975). Passive diffusion again provides the null hypothesis, and Skellam (1951) utilized this assumption; but non-linear diffusion (dependence of diffusion coefficients on the density of the diffusing species) and the effects of other species or agents can be introduced when the data so justify. Because of the analogies to other problems in continuum mechanics, continuum models of spatially distributed populations have been frequently used in the literature (see for example Aronson and Weinberger 1975; Comins and Blatt 1974; Criminale and Winter 1974; Gurney and Nisbet 1975; Gurtin 1973, 1974; Hadeler, an der Heiden, and Rothe 1974; Kierstead and Slobodkin 1953; May 1974; Montroll 1968, 1972;

Murray 1975; Noble 1974; Okubo 1975; Patten 1968; Riley 1965; Rosen 1974; Rotenberg 1972; Skellam 1951, 1973; Steele 1973, 1974);and there is of course a vast literature on the theory of diffusion-reaction equations.

For illustrative purposes, I treat briefly movement along a single linear dimension, although all of the ideas extend without complication to several dimensions. The model presented here is simple (e.g., one species), but will be extended later.

The basic assumption of all continuous diffusion models is simply that the net flux past any point is opposite to the concentration gradient of the diffusing species (u) and proportional to the gradient. The constant of proportionality $\emptyset$ can depend not only explicitly on the particular point, but also upon other variables as well, including the densities of the diffusing species and other species. Assume for this example that $\emptyset$ depends only on the point x and on μ. Then the net flux past x is $-\emptyset(x,u)\partial u/\partial x$. Note that the flux is always zero past any point where the gradient is zero. This need not be the case if tactic effects are involved, and the flux term must be appropriately modified.

Let x_1 and x_2 be any two arbitrary values of x, and $U = \int_{x_1}^{x_2} u \, dx$ the total density of organisms between x_1 and x_2. Then the rate of change of U (assuming sufficient continuity) is

$$\partial U/\partial t = \int_{x_1}^{x_2} (\partial u/\partial t)dx \quad . \tag{2}$$

However, by the flux assumption,

$$\partial U/\partial t = + \emptyset(x,u)\partial u/\partial x\big|_{x=x_2} - \emptyset(x,u) \ \partial u/\partial x\big|_{x=x_1} \quad .$$

Hence, by a fundamental theorem of the calculus,

$$\partial U/\partial t = \int_{x_1}^{x_2} (\partial/\partial x)\{\emptyset(x,u)\partial u/\partial x\}dx \quad . \tag{3}$$

Equating (2) and (3) and using the fact that x_1 and x_2 were arbitrary

12 SIMON A. LEVIN

provides the basic nonlinear diffusion equation

$$\partial u/\partial t = (\partial/\partial x)\{\emptyset(x,u)\partial u/\partial x\} \tag{4}$$

In the special case when $\emptyset(x,u)$ is a constant D^2, (4) becomes

$$\partial u/\partial t = D^2\partial^2 u/\partial x^2 . \tag{5}$$

Note that (4) and (5) include diffusion alone, and to them must be added at least the localized rates of change discussed earlier if for example growth is to be included.

For the special case where (5) applies (no net population growth), and assuming u and its first derivative vanish at $\pm\infty$, then the population size $U = \int_{-\infty}^{\infty} u\, dx$ remains constant (this is also true for (4)), since by integration by parts

$$\partial U/\partial t = \int_{-\infty}^{\infty} (\partial u/\partial t)dx = D^2 \int_{-\infty}^{\infty} (\partial^2 u/\partial x^2)dx = D^2 \int_{-\infty}^{\infty} (\partial/\partial x)(\partial u/\partial x)dx = 0 .$$

(All of this assumes, of course, that initially $U < \infty$, which is reasonable.) Further, the center of mass $M = \int_{-\infty}^{\infty} x\, u\, dx$ doesn't vary, since

$$\partial M/\partial t = D^2 \int_{-\infty}^{\infty} x(\partial^2 u/\partial x^2)dx = D^2 \int_{-\infty}^{\infty} (\partial/\partial x)\{x(\partial u/\partial x)-u\}dx = 0 .$$

(Again, initially, it is assumed that $|M| < \infty$.) Finally, as is well-known (Skellam 1951), the rate of change of the variance $V = \int_{-\infty}^{\infty} x^2 u\, dx$ is constant and, assuming the units of u are normalized so that $U = 1$, the rate is given by $2D^2$. In general,

$$\partial V/\partial t = D^2 \int_{-\infty}^{\infty} x^2(\partial^2 u/\partial x^2)dx = D^2 \int_{-\infty}^{\infty} 2x(\partial u/\partial x)dx = 2D^2 U$$

(assuming again $V < \infty$ initially). This result extends to any number of dimensions, although the constant will change. Thus the spread of variance is a logical first thing to measure in studying dispersal, and this approach has been utilized for example by Dobzhansky and Wright (1943). Any deviation from the constant rate of increase of V is evidence that dispersal is not by a process of passive diffusion.

Local uniqueness, as introduced in the previous section, is manifest through variation in the proximal interaction regions (the localized growth rates) and stochastic effects; dispersal effects operate through the last two (exchange) terms in (1). In combination, these lead to spatial patterns, to phase difference and hence to spatio-temporal patterning. The phase difference effect is thus not really independent of the other two, but is worth distinguishing.

4. Local uniqueness and spatial patterning

Exchange between patches may play a role in the establishment of spatial patterns; however, in this section I concentrate on patterns which have their origin strictly in local uniqueness. Random colonization episodes are thus viewed as deterministic or stochastic parameter variation. However, the patterns which arise through the interaction of local uniqueness and exchange with source terms are of the same type, and indeed it is completely appropriate to view the bath exchange term in (2) and (3) as a component of local uniqueness. Therefore, we henceforth ignore explicit consideration of that term as a separate entity.

The vegetational concept of the polyclimax (Whittaker 1974) refers to spatial variation in equilibrium communities due to local variations in climatic or edaphic parameters; I extend it, however, to include any spatial variation among the local equilibria which can be attributed to local uniqueness. This generalization thus includes pattern due to random founder effects in a system which admits multiple stable equilibria (Levin 1974, Sutherland 1974).

Following the development given earlier, I assume the general situation may be described in terms of the set of equations

$$du_i^\mu/dt \;=\; f_i^\mu(\underset{\sim}{u}^\mu) + g_i^\mu(\underset{\sim}{u}^1,\underset{\sim}{u}^2,\dots,\underset{\sim}{u}^m,\underset{\sim}{D}) \quad . \tag{6}$$

Here g_i^μ represents the net flux of u_i individuals into patch μ, and D a matrix of parameter values (e.g. coefficients in a Taylor expansion of the flux functions). The only assumptions made on g are those of continuity and that $g_i^\mu \geq 0$ if $u_i^\mu = 0$ and all other u-values are ≥ 0; there can be no net flux out of an empty region.

Within this formalism, then, the notion of the polyclimax is that for each μ, the system of equations

$$du_i^\mu/dt = f_i^\mu(\underset{\sim}{u}^\mu) \tag{7}$$

has a stable equilibrium. Let $\underset{\sim o}{D}$ denote a matrix of parameter values corresponding to zero dispersal $(g = 0)$. Using a particular mathematical notion of stability and the theorem proved in Appendix 1, it may be shown that the system (6) continues to have a stable equilibrium for $\underset{\sim}{D}$ sufficiently close to $\underset{\sim o}{D}$. That is, if dispersal rates are not too high (upper bounds for one special case are given in Levin 1974a), no local equilibrium can "swamp" another by migration, and the overall system pattern is one of a mosaic of equilibrium patches or "microsites" (Whittaker 1975), each slightly modified by some immigrants from nearby patches. Such pattern can occur even in homogeneous habitats through the workings of multiple steady states, and this has been suggested as a mechanism at work for example in beech-maple forests (Whittaker 1975, Smith unpubl.) and some tropical rain forests (Janzen 1970, Connell 1971). This phenomenon is not to be confused with the "fugitive" strategy to be discussed later and does not depend on differential life history strategies, although such differentiation may be expected to affect the relative densities of types. The situation is in every sense an "equilibrium" one, and would persist indefinitely if left undisturbed. Through this mechanism, space allows for the coexistence of "complete competitor" species in which competitive exclusion depends on initial advantage (Levin 1974a).

The fundamental theme which characterizes the polyclimax situation is that of competing equilibrium states at different locations, coming into conflict and competition with one another through dispersal. The essential character of the local patches is determined by local conditions and initial founder effects, and dispersal serves only to modify slightly the situation.

This assumes low rates of dispersal. In a system in which dispersal rates are higher, the conception as a "polyclimax" dominated by local autonomy gives way as the system pattern becomes more and more a blend representing the average micro-conditions rather than a discernible mosaic.

The trend towards blending may be gradual or characterized by jump discontinuities, depending upon whether the equilibria and system phase portraits change in a structurally stable way. In the case of multiple stable states, the transition will be discontinuous, with critical thresholds. For dispersal rates sufficiently high, the system is in effect "well-mixed."

The polyclimax patterns are stable in the absence of disturbance and other parametric variation. However, if extinctions are fairly common, differences among the stability characteristics of the various types of polyclimax become discernible. Those which are due strictly to microhabitat variation will, assuming the climatic and edaphic parameters remain relatively constant, be stable even in the face of extinctions and recolonizations. This still assumes a basic "equilibrium" pattern in which extinctions, though fairly common, are still infrequent enough that an individual patch is almost always in equilibrium. I shall later show how to deal with the situation when this assumption is relaxed.

The possibility of multiple successional pathways and multiple stable states introduces a random element into the history of an individual patch based on colonization episodes and other stochastic phenomena (Levin 1974a; Russell, Talbot, and Domm 1974; M.B. Davis 1974). System pattern may, however, be more predictable, representing a statistical average behavior. A mathematical approach to this is in essence to introduce a dynamic model representing transitions between various equilibrium states; in actuality such a model is implicit in (1), describing a "pseudo-steady state" behavior for (1) in which $du_i^{\mu}/dt \approx 0$ and the rates of change of the parameters are specified (see Göbber and Seelig 1975).

A variety of approaches may be devised for dealing with the colonization-

extinction balance, with the simplest assumption being that within each microhabitat type, location is unimportant; this may be modified, of course, to
include as much (and as little) geographic detail as is necessary. Since the
focus is still on an equilibrium model, it is assumed that each microhabitat
type may be in one of a finite number of equilibrium configurations, and interest focuses on the number of each microhabitat type in a particular configuration (see for example Cohen 1970, Levins and Culver 1971, Horn and Mac
Arthur 1972, Levin 1974a, Slatkin 1974).

Consider first a homogeneous habitat which has three possible states,
A_1, A_2, and the vacant state. It is important to recognize the vacant state
as an equilibrium; the assumption that the dynamic time scale in (1) is rapid
makes no evaluation of the relative rates of extinction of patches as compared
with rates of recolonization of empty patches.

A_1 and A_2 may be characterized each by a single species, e.g., sugar
maple (A_1) and birch (A_2), or by complexes of species. Let P_1 = the number
of patches in state A_1, P_2 = number in state A_2, and K = total number of
patches, so that $K - P_1 - P_2$ are vacant. K is assumed constant, although
it too could be allowed to fluctuate over time. A basic model then might take
the form

$$dP_1/dt = F_1(P_1, P_2, t) ,$$

$$dP_2/dt = F_2(P_1, P_2, t) ,$$

(8)

where the time dependence represents seasonal or other temporal variation in
regional characteristics, immigration rates, etc. Such equations can allow
for competitive extinction caused by invasions from other patches through the
dependence say of F_1 on P_2 or of F_2 on P_1, and for recolonizations
either from other patches or from a bath. The systems (1) and (8) provide two
of a hierarchy of models layered by time scales. If parametric variation is
on a yet slower time scale in (8), then (8) may be viewed as autonomous and
one may search for stationary or periodic attracting states in (8) which track

parametric variation, and this dissects the dynamics even further temporally.

Yet further dissection   occurs if one generalizes the notion of the pseudo-steady state to include stages which succeed one another (successional states), but are long-lived as regards transitory dynamics. In this approach, the spatial patterning is more a matter of phase difference than local uniqueness, and is left until the next section.

5. Phase difference and spatio-temporal patterning

The patterns discussed in the preceding section are due to local differentiation, and may or may not vary over time. In this section, I concentrate on patterns which require a difference in timing as an ingredient, and thus must show temporal variation. The two kinds of patterns intergrade, but I attempt to maintain the distinction by initially assuming a homogeneous environment (no local uniqueness).

The successional process alluded to at the end of the preceding section may of course have more than two occupied stages, and a hierarchy of models are discussed in Levin (1977). As one provides more details of the successional process, one sees succession less and less as a progression of discrete stages and more and more as a temporal continuum; the models developed assume a form similar to that employed by Levin and Paine (1974, 1975).

The work described in Levin and Paine (1974, 1975) was developed to model the marine rocky intertidal fauna and flora, but the approach extends to any system in which disturbance is a random localized occurrence interrupting the successional process or destroying the climax, renewing the temporal sequence. Fires and blowdowns in forests (Taylor 1973, Heinselman 1973, Poore 1964, Horn 1975, 1976) provide other examples.

It is assumed in such models that the system may be characterized by the distributional properties (as regards age, size, and other characteristics), of the patches ("islands") which disturbance creates, and by the biological properties of the individual patches. The approach is to develop a "continuity" equation similar to that utilized by von Foerster (1959) for cell populations.

Let $n(t, a, \vec{\xi})$ = the density function for patches of age a and characteristics (e.g., size) $\vec{\xi}$ at time t. Further, assume that the mean value of
$d\xi_k/dt$ is given by $g_k(t, a, \vec{\xi})$. In the simplest case, g_k is the growth rate
of a patch and will be negative if patches shrink. Finally, let $\mu(t, a, \vec{\xi})$ be
the death (extinction) rate of patches of age a, and characteristics $\vec{\xi}$ at
time t. Then $n(t, a, \vec{\xi})$ is governed by the equation

$$\frac{\partial n}{\partial t} + \frac{\partial n}{\partial a} + \Sigma_k \frac{\partial(g_k n)}{\partial \xi_k} = -\mu n \quad ,$$

and may be solved in closed form given the initial distribution $n(0, a, \vec{\xi})$ and
the rate of new patch formation $n(t, 0, \vec{\xi})$. More generally, however, $n(t, 0, \vec{\xi})$
may not be given explicitly but may be specified by an integral-type feedback
relationship, and the solution process is more difficult (Oster and Takahashi
1974, Rubinow 1973).

Similarly, if there is noise in the data so that not only the mean but
also the variance of $d\xi_k/dt$ need be considered, a different model results
(Levin and Paine 1975, Oster and Takahashi 1974, Rubinow 1973), a second-order
parabolic equation. In practice, even this approach may be substantially more
deterministic than Nature.

Models of this type are designed to relate patterns of environmental
heterogeneity to patterns of disturbance by developing relations between disturbance and the patch or insular structure of the environment. In turn, this
island structure determines patterns of species diversity; island biogeograph-
ical studies (MacArthur and Wilson 1967, Simberloff and Wilson 1969, 1970,
Root 1973) are designed to explore this relationship.

The model described in this section is only a crude first attempt, and
the reader is directed to Levin and Paine (1975) for discussion of some exten-
sions.

6. Dispersal and patterning

Localized disturbance plus dispersal of new colonists can create patterns in species distributions, and this was the topic of the preceding section. In that section, pattern was the result of disturbance unrelated to dispersal, and dispersal was simply the means by which species exploited environmental pattern.

In this section, focus is upon patterns which arise endogenously due to the interaction of dispersal and local dynamics. To make the matter clearer, we consider the system as closed, sealed off from extrinsic factors. Patterns may include a temporal component or be time independent. The former are of great importance (Huffaker 1958, Hutchinson 1959; see also Kopell and Howard 1975, Othmer and Scriven 1971, 1973), and bear a close relation to those discussed in the previous section with the source area included as part of the system. The mathematical problems are difficult, and in an ecological setting the treatment of them to date has been heuristic (Maynard Smith 1974).

In this section we concentrate on another phenomenon, the development of time-independent spatial pattern through the joint influence of dispersal and population interactions. Steele (1973, 1974) has suggested this as a mechanism for the formation of patchiness in the plankton; and in another context, the work of Turing (1952) has given rise to a broad literature on the development of biological pattern (Gierer and Meinhardt 1972, 1975, Meinhardt and Gierer 1974, Keller and Segel 1971).

For simplicity, we consider two species: a predator E and its prey V, and begin from the governing equations (Segel and Levin 1975)

$$\frac{dV}{dt} = Vf(V) - P(E,V) \quad , \quad \frac{dE}{dt} = Eg(E) + \xi P(E,V) \quad . \tag{9}$$

This system may or may not admit a stable equilibrium (Segel and Levin 1975); and even if it does not, a stable limit cycle will exist for most cases of biological interest (May 1972). Further, Steele has suggested that these unstable equilibria may contribute to planktonic patch formation when spatial diffusion terms are added.

Our interest (Segel and Jackson 1972, Levin 1974a, Segel and Levin 1975)

and that of Okubo (1975) has centered on another phenomenon: the destabiliza-
tion of the uniform spatial distribution through the action of diffusion, and
the subsequent attainment of a new spatial pattern. We consider here only
passive diffusion, but the approach extends to more general movement regimes.
To examine this question, we replace (9) by the system of partial differential
equations (Segel and Levin 1975)

$$\partial V/\partial t = V(K-\alpha V -\beta E) + \nabla\cdot(\mu\nabla V) \ ,$$
$$\partial E/\partial t = E(-L+\gamma V-\delta E) + \nabla\cdot(\nu\nabla E)$$

$$(10)$$

in the continuous spatial case, and in the discrete case by

$$dV_i/dt = V_i(K-\alpha V_i -\beta E_i) + \sum_j d_{ij}(V_j-V_i) \ ,$$
$$dE_i/dt = E_i(-L+\gamma V_i-\delta E_i) + \sum_j d'_{ij}(E_j-E_i)$$

$$(11)$$

where ∇ is the gradient operator with respect to space in the continuous
case, and the subscripts i, j in the discrete case refer to patch number.
The replacement of the local dynamics by a quadratic representation is no
restriction, since interest will be focused on behavior in the neighborhood
of the uniform equilibrium. It is assumed that $K > 0$, $\beta > 0$, and $\gamma > 0$,
since the interaction is predator-prey. Further, for simplicity in calculation we consider here the case $L = 0$ and assume that the diffusion coefficients are constants.

For convenience, the systems (10) and (11) are non-dimensionalized (Segel
and Levin 1975), which transforms the basic equations into

$$\partial\bar{v}/\partial\bar{t} = (1+\bar{k}\bar{v})\bar{v} - a\bar{e}\bar{v} + \bar{\nabla}^2\bar{v} \ ,$$
$$\partial\bar{e}/\partial\bar{t} = \bar{e}\bar{v} - \bar{e}^2 + \theta^2\bar{\nabla}^2\bar{e}$$

$$(12)$$

for the continuous case, and

$$d\bar{v}_i/d\bar{t} = (1+\bar{k}\bar{v}_i)\bar{v}_i - a\bar{e}_i\bar{v}_i + \sum \bar{d}_{ij}(\bar{v}_j-\bar{v}_i) \ ,$$
$$d\bar{e}_i/d\bar{t} = \bar{e}_i\bar{v}_i - \bar{e}_i^2 + \theta^2\sum \bar{d}_{ij}(\bar{e}_j-\bar{e}_i)$$

$$(13)$$

in the discrete.

It is assumed that $a > \bar{k}$; otherwise no uniform steady state exists, stable or otherwise. I shall discuss later the constraints on $\bar{k}$. θ^2 is a very important parameter, the <u>diffusivity</u> <u>ratio</u>, and relates predator diffusive ability to prey.

Both sets of equations have a non-trivial <u>steady</u> <u>state</u> <u>spatially</u> <u>uniform</u> solution given by

$$\bar{e} = \bar{v} = p^{-2}$$

in the continuous and

$$\bar{e}_i = \bar{v}_i = p^{-2}$$

in the discrete. Here $p^2 = a - \bar{k} > 0$.

The stability of the uniform state is investigated by writing equations for the deviations (e, v) from steady state densities (p^{-2}, p^{-2}), and these equations take the form

$$-(\partial \underset{\sim}{V}/\partial \bar{t}) + \underset{\sim\theta}{M}\underset{\sim}{V} = \underset{\sim}{N}(\underset{\sim}{V}) \tag{14}$$

in the continuous case, and

$$-(d\underset{\sim}{V}/d\bar{t}) + \underset{\sim\theta}{M}^{*}\underset{\sim}{V} = \underset{\sim}{N}(\underset{\sim}{V}) \tag{15}$$

in the discrete. Here, in the continuous case

$$\underset{\sim}{V} = (v,e)^T \ , \quad \underset{\sim}{N}(\underset{\sim}{V}) = (aev - \bar{k}v^2, e^2 - ev)^T \ ,$$

$$\text{and} \quad \underset{\sim\theta}{M} = \begin{pmatrix} \bar{k}p^{-2} + \triangledown^2 & -ap^{-2} \\ \\ p^{-2} & -p^{-2} + \theta^2\triangledown^2 \end{pmatrix} .$$

In the discrete,

$$\underset{\sim}{V} = (v_1, e_1, v_2, e_2, \ldots, v_n, e_n)^T \ ,$$

$$N(\underset{\sim}{V}) = (ae_1v_1 - \bar{k}v_1^2, e_1^2 - e_1v_1, ae_2v_2 - \bar{k}v_2^2, e_2^2 - e_2v_2, \ldots, ae_nv_n - \bar{k}v_n^2, e_n^2 - e_nv_n)^T \ ,$$

and

$$\underset{\sim}{M}{}_{\theta}^{*} = p^{-2}\underset{\sim}{I}_n \otimes \begin{pmatrix} \bar{k} & -a \\ 1 & -1 \end{pmatrix} + \triangle \otimes \begin{pmatrix} 1 & 0 \\ 0 & \theta^2 \end{pmatrix}$$

in which $\otimes$ denotes tensor product, I_n is the $n \times n$ identity matrix, and $\triangle$ is the "structural matrix" (Othmer and Scriven 1971) with components $\bar{d}_{ij}$.

I consider first the continuous case; the results in the discrete case are basically similar, with some exceptions. Stability of the equilibrium is investigated with respect to perturbations of the form

$$\underset{\sim}{V} = \underset{\sim}{C} \exp(\sigma t) \exp(i\underset{\sim}{w} \cdot \underset{\sim}{x}) \quad . \tag{16}$$

If $\bar{k} \leq 0$, the uniform equilibrium is stable to any such disturbance; if $\bar{k} > 1$, it is unstable even to those of infinite wavelength $(q = |\underset{\sim}{w}| = 0)$. If $0 < \bar{k} < 1$, a more peculiar situation occurs. The equilibrium is stable to perturbations of infinite wavelength (i.e., the equilibrium is stable if no diffusion occurs); but its stability in the face of other perturbations depends on the ratio θ. In particular, if $\theta = 1$ (predator and prey diffuse equally), it is stable to any perturbation; and this is in fact true as long as $\theta < 1/(\sqrt{a} - \sqrt{a-\bar{k}})$. [Note that since $a > \bar{k}$ and $1 > \bar{k} > 0$, it follows that $\sqrt{a} - \sqrt{a-\bar{k}} < 1$.] If, however, $\theta > \theta_c = 1/(\sqrt{a} - \sqrt{a-\bar{k}})$, perturbations of wave number $q = (\sqrt{a} - \sqrt{a-\bar{k}})/\sqrt{a-\bar{k}}$ will destabilize the uniform equilibrium. This diffusive instability is analogous to that discussed by Turing (1952), and by many writers since, as a source of patchiness. However, discussion usually is restricted to linear analysis. In Segel and Levin (1975), the non-linear analysis is pursued by application of a combined successive-approximations multiple-scale approach. For θ close to θ_c, it is shown that a new non-uniform steady state appears, with clumping patterns in both species.

The case $\bar{k} > 0$ corresponds to an auto-catalytic (Allee-type) effect in the prey, and may be justified on various grounds (Segel and Levin 1975).

The analysis shows that in a homogeneous environmental continuum, random diffusion can destabilize an otherwise stable uniform predator-prey distribution provided that predator dispersal ability is sufficiently greater than

that of the prey; but that as the perturbations grow, non-linear effects lead
to a new steady but spatially non-uniform pattern of species distributions.
In the discrete (patchy) spatial case, the results are similar except that
a "sub-critical instability" can arise in which the amplitude of the perturbation may become a consideration. The general conclusion remains valid:
"Patchiness in species distributions may arise even in a homogeneous environment through species interactions." Segel and Levin (1975), Okubo (1974) and
Levin and Segel (1975) have suggested this as a possible cause of the patchy
distribution of marine phytoplankton, and Steele (1973, 1974a,b) has discussed the analogous phenomenon for the unstable case $\bar{k} > 1$.

The case $\bar{k} > 0$ corresponds to a "destabilizing functional response" in
the sense of Oaten and Murdoch (1975); but as shown above and emphasized in
Levin (1975), even if the functional response destabilizes the uniform equilibrium, a new stable non-uniform one may obtain.

7. Summary

In a variety of ways the spatial extent of environment, largely ignored
in mathematical modelling, may fundamentally alter conclusions of ecological
theory and operate to increase species diversity. As pointed out in Levin
(1974a), this arises due to heterogeneity of environment but such heterogeneity
can arise in an otherwise homogeneous environment due to random events magni-
fied by species interactions. In this way, "homogeneous environments may become heterogeneous and heterogeneous environments even more so" (Levin 1974).

Space complicates ecological interactions by two fundamental mechanisms:
it allows for non-uniform patterns of environment and population density, and
for movement (dispersal or migration) from one location to another. The factors which impinge are grossly categorized as due to local uniqueness, phase
difference, and dispersal; and diversity is increased through combinations of
these.

Local uniqueness includes not only spatial parametric variation, but also
the divergent local regimes which result from random events coupled with

multiple basins of attraction. As explored in Levin (1974a), initial colonization patterns create a "founder effect" so that species which are unable to
invade areas may survive by early establishment and subsequent resistance to
invasion by competitors. Understanding of such patterns requires an approach
to the problem through a slower time scale, that of colonization and extinction.

If disturbance (extinction) is common enough that the latter time scale
is not really a slower one, then the differences in maturity (phase difference)
of local environments become important. Such a situation obtains especially
where a species is sessile for most of its adult life, such as in the marine
rocky intertidal and in forests. A model is described, following Levin and
Paine (1974, 1975) to treat such situations.

Dispersal mitigated through a source area (e.g., an oceanic planktonic
bath)is obviously important as an extrinsic contributor to diversity. However,
even in closed systems dispersal between local areas can be an important factor
in the development of spatial or spatio-temporal patterns. In the preceding
section of this paper, a particular instance is considered, involving a basic
predator-prey model; and this leads to speculations with regard to herbivores
and phytoplankton. Random dispersal is modelled by diffusion-like terms in
both continuous and discrete environments. In the presence of a destabilizing
functional response, such passive diffusion can destabilize an otherwise stable
homogeneous equilibrium if predator dispersal is sufficiently high, but this
may be succeeded by a new spatially non-uniform pattern.

Appendix 1

A principal result of Section 4 rests on the following theorem (Levin
1974a).

THEOREM: Assume that $F_i^\mu(\underset{\sim}{U}, \underset{\sim}{D}) \geq 0$ when $u_i^\mu = 0$ and $\underset{\sim}{U} \geq 0$, and that

the system $du_i^\mu/dt = F_i^\mu(\underset{\sim}{U}, \underset{\sim}{D}_o)$ has a stable equilibrium (in the

sense that the eigenvalues of the Jacobian matrix have negative real parts) at $\underset{\sim}{U} = \underset{\sim o}{U} \geq 0$. Then for $\underset{\sim}{D}$ sufficiently close

to $\underset{\sim o}{D}$, the system $du_i^\mu/dt = F_i^\mu(\underset{\sim}{U},\underset{\sim}{D})$ has a stable equilibrium

at some point $\underset{\sim D}{U} \geq 0$, where $\underset{\sim D}{U}$ tends to $\underset{\sim o}{U}$ as $\underset{\sim}{D}$ tends

to $\underset{\sim o}{D}$.

Here $\underset{\sim}{U} = (\underset{\sim}{u}^1,\underset{\sim}{u}^2,\ldots,\underset{\sim}{u}^m)$ where $\underset{\sim}{u}^\mu = (u_i^\mu,\ldots,u_n^\mu)$. The notation $\underset{\sim}{U} \geq 0$

means that all components of $\underset{\sim}{U}$ are non-negative. The quantity F_i^μ is as-

sumed continuously differentiable in $\underset{\sim}{U}$ and $\underset{\sim}{D}$.

A discrete time version of this theorem was derived by Karlin and McGre-

gor (1972); the continuous version was given in Levin (1974a) without the proof

herein provided.

The theorem is first cast in a more general framework, and the proof pro-

vided for that version. The notion of stability employed is that the lineari-

zation matrix of the system about its equilibrium admits only eigenvalues with

negative real parts. This is a sufficient condition for asymptotic stability,

but the linearization matrix in an asymptotically stable system might admit

some eigenvalues with zero real part. In this borderline case, the results of

this Appendix do not apply.

THEOREM: Let $\underset{\sim}{y} = (y_1,\ldots,y_p), \underset{\sim}{z} = (z_1,\ldots,z_q)$; and for $j = 1,\ldots,p$,

let $G_j(\underset{\sim}{y},\underset{\sim}{z})$ be a continuously differentiable function of

$\underset{\sim}{y}$ and $\underset{\sim}{z}$ and satisfy the inequality $G_j(\underset{\sim}{y},\underset{\sim}{z}) \geq 0$ provided

$y_j = 0$ and $\underset{\sim}{y} \geq 0$. Assume further that the system of differ-

ential equations

$$\frac{dy_j}{dt} = G_j(\underset{\sim}{y},\underset{\sim}{z}^\circ)$$

has a stable equilibrium at $\underset{\sim}{y} = \underset{\sim}{y}^\circ \geq 0$. Then for $\underset{\sim}{z}$ suf-

ficiently close to $\underset{\sim}{z}^\circ$, the system

$$\frac{dy_j}{dt} = G_j(\underset{\sim}{y},\underset{\sim}{z})$$

also has a stable equilibrium $\underset{\sim}{y}(\underset{\sim}{z}) \geq 0$, and $\underset{\sim}{y}(\underset{\sim}{z})$ tends to

$$\underset{\sim}{y}^{\circ} \quad \text{as} \quad \underset{\sim}{z} \quad \text{tends to} \quad \underset{\sim}{z}^{\circ}.$$

Proof: Since "stability" guarantees the non-vanishing of the Jacobian matrix, the existence of the equilibrium and continuous dependence on $\underset{\sim}{z}$ follows from the implicit function theorem. Retention of the stability character follows from the continuous dependence of the eigenvalues upon $\underset{\sim}{z}$. The only difficult part of the proof is thus the demonstration that $\underset{\sim}{y}(\underset{\sim}{z}) \geq 0$ for $\underset{\sim}{z}$ close to $\underset{\sim_{o}}{z}$.

If $\underset{\sim}{y}^{\circ} > 0$, the above assertion is trivial, and follows from continuous dependence. Therefore assume, without loss of generality, that there is an m such that

$$y_1^{\circ} = y_2^{\circ} = \ldots = y_m^{\circ} = 0;$$

and that $y_j^{\circ} > 0$ for $j > m$. This may always be achieved by a permutation. The trivial case $m = p$ is admissible.

To begin, some properties of the Jacobian matrix $J = ((\partial G_j / \partial y_k))$ are derived; in particular at the equilibrium (where we use the notation $J = J_o$). At equilibrium, by definition, $G_j = 0$ for all j. Further, if $j \leq m$, then $\partial G_j / \partial y_k \geq 0$ for any $k \neq j$; and in fact $\partial G_j / \partial y_k = 0$ if $k > m$. This follows from the condition $G_j \geq 0$ when $y_j = 0$ and $\underset{\sim}{y} \geq 0$. Thus J_o takes the form

$$J_o = \begin{bmatrix} \partial G_1/\partial y_1 & \cdots & \partial G_1/\partial y_m & 0 & \cdots & 0 \\ \vdots & & \vdots & \vdots & & \vdots \\ \partial G_m/\partial y_1 & \cdots & \partial G_m/\partial y_m & 0 & \cdots & 0 \\ \partial G_{m+1}/\partial y_1 & \cdots & \partial G_{m+1}/\partial y_m & \partial G_{m+1}/\partial y_{m+1} & \cdots & \partial G_{m+1}/\partial y_p \\ \vdots & & \vdots & \vdots & & \vdots \\ \partial G_p/\partial y_1 & \cdots & \partial G_p/\partial y_m & \partial G_p/\partial y_{m+1} & \cdots & \partial G_p/\partial y_p \end{bmatrix}$$

By assumption, all of the eigenvalues of this matrix have negative real part; and hence the same must be true for the restriction

$$A_o = \begin{pmatrix} \partial G_1/\partial y_1 & \cdots & \partial G_1/\partial y_m \\ \vdots & & \vdots \\ \partial G_m/\partial y_1 & \cdots & \partial G_m/\partial y_m \end{pmatrix}$$

As shown earlier, this matrix has non-negative off-diagonal entries; and hence (Gantmakher 1959) every principal submatrix B_o is a stability matrix. It is shown in Levin (1974b) that for such a matrix the system $B_o \underset{\sim}{x} \geq 0, \underset{\sim}{x} > 0$ has no non-trivial solution. In other words, if $\underset{\sim}{x} \geq 0$ and $\underset{\sim}{x} \neq 0$, at least one component of $B_o \underset{\sim}{x}$ is strictly negative. From the compactness of the set $\{\underset{\sim}{x}: \underset{\sim}{x} \geq 0, \ ||\underset{\sim}{x}|| = 1\}$ and continuity, the comparable restriction B of the Jacobian matrix J evaluated at points sufficiently close to $(\underset{\sim}{y}^\circ, \underset{\sim}{z}^\circ)$ satisfies the same property. That is, if $(\underset{\sim}{y}, \underset{\sim}{z})$ is sufficiently close to $(\underset{\sim}{y}^\circ, \underset{\sim}{z}^\circ)$ (say within Euclidean distance δ) and if B is the appropriate restriction of J evaluated at $(\underset{\sim}{y}, \underset{\sim}{z})$, then $B\underset{\sim}{x} \geq 0, \ \underset{\sim}{x} \geq 0$ has no non-trivial solution. The proof proceeds first for $\underset{\sim}{x}$ in the set $\{\underset{\sim}{x}: \underset{\sim}{x} \geq 0, \ ||\underset{\sim}{x}|| = 1\}$, and then by extension to $\{\underset{\sim}{x}: \underset{\sim}{x} \geq 0, \ \underset{\sim}{x} \neq 0\}$. Although this result will be applied at points other than the equilibria $(\underset{\sim}{y}(\underset{\sim}{z}), \underset{\sim}{z})$, it suffices that $\underset{\sim}{z}$ be chosen close enough to $\underset{\sim}{z}^\circ$ that (utilizing continuous dependence) $(\underset{\sim}{y}(\underset{\sim}{z}), \underset{\sim}{z})$ is within δ of $(\underset{\sim}{y}^\circ, \underset{\sim}{z}^\circ)$. Application to non-equilibrium points will always involve points $(\underset{\sim}{y}, \underset{\sim}{z})$ closer to $(\underset{\sim}{y}^\circ, \underset{\sim}{z}^\circ)$ than is $(\underset{\sim}{y}(\underset{\sim}{z}), \underset{\sim}{z})$.

Consider now the equilibrium point $(\underset{\sim}{y}^\circ, \underset{\sim}{z}^\circ)$, and the new equilibrium $(\underset{\sim}{y}(\underset{\sim}{z}), \underset{\sim}{z})$ nearby. For each j, G_j vanishes at both points. Define $\underset{\sim}{y}^*$ to be identical to $\underset{\sim}{y}(\underset{\sim}{z})$ for each positive component of $\underset{\sim}{y}(\underset{\sim}{z})$ and let the other components of $\underset{\sim}{y}^*$ be zero. Without loss of generality, assume that $\underset{\sim}{z}$ is close enough to $\underset{\sim}{z}^\circ$ that by continuity, every component which is positive in $\underset{\sim}{y}^\circ$ remains positive in $\underset{\sim}{y}(\underset{\sim}{z})$; and that exactly the first r components of $\underset{\sim}{y}^*$ are zero. Then $r \leq m$; and furthermore $y_j(\underset{\sim}{z}) > 0$ for $j > r$, and $y_j(\underset{\sim}{z}) \leq 0$ for $j \leq r$. (Note that $||(\underset{\sim}{y}^*, \underset{\sim}{z}) - (\underset{\sim}{y}^\circ, \underset{\sim}{z}^\circ)|| \leq ||(\underset{\sim}{y}(\underset{\sim}{z}), \underset{\sim}{z}) - (\underset{\sim}{y}^\circ, \underset{\sim}{z}^\circ)||.$) Then, for $j \leq r$, $G_j(\underset{\sim}{y}^*, \underset{\sim}{z}) \geq 0$. Using the mean-value theorem, one obtains (for $j \leq r$)

$$0 \leq G_j(\underset{\sim}{\chi}^*, \underset{\sim}{z}) - G_j(\underset{\sim}{\chi}(\underset{\sim}{z}), \underset{\sim}{z}) = \sum_{k=1}^{r} (-y_k(\underset{\sim}{z})) \partial G_j / \partial y_k \quad,$$

where partial differentiation is taken at some intermediate point.

Utilizing the result derived earlier, with B the $r \times r$ leading principal submatrix of J, we see that since $-y_k(\underset{\sim}{z}) \geq 0$ for $k \leq r$, $-y_k(\underset{\sim}{z})$ must equal zero for every k. Thus $\underset{\sim}{y}(\underset{\sim}{z}) \geq 0$, which was to be proved.

In Levin (1974) some specific examples are considered. Thus, for example, consider the system (of two species competing in two patches)

$$du_1^\mu / dt = u_1^\mu (R - au_1^\mu - bu_2^\mu) + D(u_1^\nu - u_1^\mu) \quad,$$

$$du_2^\mu / dt = u_2^\mu (R - bu_1^\mu - au_2^\mu) + D(u_2^\nu - u_2^\mu) \quad,$$

where $\mu, \nu = 1, 2$ and $\mu \neq \nu$, and where $0 < a < b$ (interspecific competition exceeds intraspecific). With $D = 0$ the two species being modelled can not coexist in any patch; the only stable equilibria are the homogeneous one

$$u_1^\mu = u_1^\nu = R/a, \ u_2^\mu = u_2^\nu = 0$$

and its symmetric analogue, and the completely disjoint inhomogeneous ones

$$u_1^\mu = u_2^\nu = R/a, \ u_2^\mu = u_1^\nu = 0$$

and its symmetric analogue. However, when D is non-zero and not too large, both species may be found in both patches; two stable non-trivial equilibria are given by

$$u_1^\mu = u_2^\nu = \frac{R-2D}{2a} \pm \frac{1}{2a} \sqrt{(R-2D)(R-2D\frac{b+a}{b-a})} \quad,$$

$$u_2^\mu = u_1^\nu = \frac{R-2D}{2a} \mp \frac{1}{2a} \sqrt{(R-2D)(R-2D\frac{b+a}{b-a})} \quad.$$

These equilibria exist for $0 \leq D \leq \frac{R}{2}\frac{b-a}{b+a}$, and are stable for

$$0 \leq D \leq \frac{R}{2}\frac{b-a}{2b+a} \quad.$$

The global behavior of the system, in particular for

$$\frac{R}{2}\frac{b-a}{2b+a} < D \leq \frac{R}{2}\frac{b-a}{b+a} \quad ,$$

remains an unsettled question. In the locally stable case, some sufficient conditions have been derived for global stability by the theory of differential inequalities (S.A. Levin and L.E. Payne, unpublished); but even in this case, the problem is not completely solved.

CORNELL UNIVERSITY

Acknowledgments

This research has been supported by NSF Grant GP33031. Earlier versions of this manuscript have been read and commented upon by Dan Cohen, Alan Hastings, Robert T. Paine, Richard Root, George White, and Robert H. Whittaker. Their candid comments have been invaluable.

BIBLIOGRAPHY

Adler, J. 1966. Chemotaxis in bacteria. Science 153:708-716.

Adler, J. 1969. Chemoreception in bacteria. Science 166:1588-1597.

Allen, J.C. 1975. Mathematical models of species interactions in time and space. Amer. Natur. 109:319-342.

Aronson, D.G. and H.F. Weinberger. 1975. Nonlinear diffusion in population genetics, combustion, and nerve propagation. In Proc. Tulane Program in Partial Differential Equations, Lecture Notes in Mathematics, Springer-Verlag, Heidelberg.

Ayala, F. 1971. Competition between species: frequency dependence. Science 171:820-824.

Charlesworth, B. 1970. Natural selection in age-structured populations. This volume.

Chet, I., Y. Henis and R. Mitchell. 1973. Effect of biogenic amines and cannabinoids on bacterial chemotaxis. J. Bacteriol. 115:1215-1218.

Chewning, W. 1975. Migratory effects in predator-prey models. Math. Biosci. 23:253-262.

Clements, F.E. 1928. Plant succession and indicators. H.W. Wilson Co., New York. 453 pp.

Cohen, J.E. 1970. A Markov contingency table model for replicated Lotka-Volterra systems near equilibrium. Amer. Natur. 104:547-559.

Comins, H.N. and W.E. Blatt. 1974. Prey-predator models in spatially hetero-
    geneous environments. J. Theor. Biol. 48:75-83.

Commito, J.A., and D.I. Rubenstein. 1975. Patch utilization and resource
    limitation in disturbance-regulated communities. MS.

Connell, J.H. 1971. On the role of natural enemies in preventing competitive
    exclusion in some marine animals and in rain forest trees. Pages 298-312
    in P.J. Den Boer and G.R. Gradwell (Eds.), Dynamics of Populations. Proc.
    Adv. Study Inst. on Dynamics of Numbers and Populations. Oosterbeek,
    Netherlands. Sept. 7-18, 1970. Wageningen, Centre for Agricultural Pub-
    lishing and Documentation.

Criminale, W.O., Jr. and D.F. Winter. 1974. The stability of steady-state
    depth distributions of marine phytoplankton. Amer. Natur. 108:679-687.

Davis, M.B. 1974. Pleistocene biogeography of temperate deciduous forests.MS.
Demetrius, L. 1975. Age-structured models in population biology. MS.
Den Boer, P.J. 1968. Spreading of risk and stabilization of animal numbers.
    Acta biotheor. Leiden XVIII:165-194.

Den Boer, P.J. 1971. Stabilization of animal numbers and the heterogeneity
    of the environment: the problem of the persistence of sparse populations.
    Pages 77-97 in P.J. Den Boer and G.R. Gradwell (Eds.), Dynamics of Popula-
    tions. Proc. Adv. Study Inst. on Dynamics of Numbers and Populations.
    Oosterbeek, Netherlands. Sept. 7-18, 1970. Wageningen, Centre for Agri-
    cultural Publishing and Documentation.

Dobzhansky, Th., and S. Wright. 1943. Genetics of natural populations. X.
    Dispersion rates in Drosophila pseudoobscura. Genetics 28:304-340.

Dodd, A.P. 1940. The biological campaign against prickly pear. Communica-
    tions Prickly Pear Board, Brisbane, Queensland.

Dodd, A.P. 1959. The biological control of prickly pear in Australia. Pages
    565-577 in A. Keast, R.L. Crocker, and C.S. Christian (Eds.), Biogeography
    and Ecology in Australia. Mongr. Biol. 8.

Gadgil, M. 1971. Dispersal: population consequences and evolution. Ecology
    52:253-260.

Gantmakher, F.R. 1959. The theory of matrices. Vol. 1. Chelsea, New York.
    374 + x pp.

Gierer, A. and H. Meinhardt. 1972. The theory of biological pattern formation.
    Kybernetik 12:30-39.

Gierer, A. and H. Meinhardt. 1974. Biological pattern formation involving
    lateral inhibition. Pages 163-184 in S.A. Levin (Ed.), Some Mathematical
    Questions in Biology, VI. Lectures on Mathematics in the Life Sciences,
    Vol. 7. American Mathematical Society, Providence.

Göbber, F. and F.F. Seelig, 1975. Conditions for the application of the
    steady-state approximation to systems of differential equations. J. Math.
    Biol. 2:79-86.

Gurney, W.S.C. and R.M. Nisbet. The regulation of unhomogeneous populations.
    J. Theor. Biol. 52:441-457.

Gurtin, M.E. 1973. A system of equations for age-dependent population diffusion. J. Theor. Biol. 40:389-392.

Gurtin, M.E. 1974. Some mathematical models for population dynamics that lead to segregation. Quart. Appl. Math. 32:1-9.

Hadeler, K.P., U. an der Heiden, and F. Rothe. 1974. Nonhomogeneous spatial distributions of populations. J. Math. Biol. 1:165-176.

Heinselman, M.L. 1973. Fire in the virgin forests of the Boundary Waters Canoe Area, Minnesota. J. Quaternary Res. 3:329-382.

Horn, H.S. 1974. The ecology of secondary succession. Ann. Rev. Ecol. Syst. 5:25-37.

Horn, H.S. 1975. Forest succession. Sci. Amer. 232:90-98.

Horn, H.S. 1976. Markovian properties of forest succession. In J. Diamond and M.L. Cody (eds.), Ecology of Communities. Harvard University Press.

Horn, H.S. and R.H. MacArthur. 1972. Competition among fugitive species in a harlequin environment. Ecology 53:749-752.

Huffaker, C.B. 1958. Experimental studies on predation: dispersion factors and predator-prey oscillations. Hilgardia 27:343-383.

Hutchinson, G.E. 1959. Homage to Santa Rosalia, or why are there so many kinds of animals? Amer. Natur. 93:145-159.

Janzen, D.H. 1970. Herbivores and the number of tree species in tropical forests. Amer. Natur. 104:501-528.

Karlin, S. and J. McGregor. 1972. Polymorphisms for genetic and ecological systems with weak coupling. Theor. Pop. Biol. 3:210-238.

Keller, E.F. and L.A. Segel. 1970. Initiation of slime mold aggregation viewed as an instability. J. Theor. Biol. 26:399-415.

Kierstead, H., and L.B. Slobodkin. 1953. The size of water masses containing plankton blooms. J. Mar. Res. 12:141-147.

Kopell, N. and L.N. Howard. 1974. Pattern formation in the Belousov reaction. Pages 201-217 in S.A. Levin (ed.), Some Mathematical Questions in Biology. VI. Lectures on Mathematics in the Life Sciences, Vol. 7. American Mathematical Society, Providence.

Leon, J.A. 1974. Selection in contexts of interspecific competition. Amer. Natur. 108:739-757.

Levin, S.A. 1970. Community equilibria and stability, and an extension of the competitive exclusion principle. Amer. Natur. 104:413-423.

Levin, S.A. 1972. A mathematical analysis of the genetic feedback mechanism. Amer. Natur. 106:145-164.

Levin, S.A. 1973. Erratum. Amer. Natur. 107:320.

Levin, S.A. 1974a. Dispersion and population interactions. Amer. Natur. 108: 207-228.

Levin, S.A. 1974b. Stability matrices and the solubility of certain systems of linear inequalities. Linear and Multilinear Algebra 2:253-255.

Levin, S.A. 1975. A more functional response to predator-prey stability.MS.

Levin, S.A. 1977. Population dynamics in heterogeneous environments. Ann. Rev. of Ecology and Systematics VII (to appear).

Levin S.A. and R.T. Paine. 1974. Disturbance, patch formation, and community structure. Proc. Nat. Acad. Sci. 71:2744-2747.

Levin S.A. and R.T. Paine. 1975. The role of disturbance in models of community structure. Pages 56-67 in S.A. Levin (Ed.) Ecosystem Analysis and Prediction, SIAM, Philadelphia, Pa.

Levin, S.A. and L.A. Segel. 1975. An hypothesis for the origin of planktonic patchiness. MS.

Levin S.A. and J.D. Udovic. 1975. A mathematical model of coevolving populations. M S.

Levins, R. and D. Culver. 1971. Regional coexistence of species and competition between rare species. Proc. Nat. Acad. Sci. 68:1246-1248.

Lewontin, R.C. 1969. The Meaning of Stability. Pages 13-24 in Diversity and Stability in Ecological Systems. Brookhaven Symposium in Biology 22.

Lin, C.C. and L.A. Segel. 1974. Mathematics Applied to Deterministic Problems in the Natural Sciences. MacMillan, New York. 604 + x i x pp.

Loucks, O.L. 1970. Evolution of diversity, efficiency, and community stability. Amer. Zool. 10:17-25.

Luckinbill, L.S. 1973. Coexistence in laboratory populations of Paramecium aurelia and its predator Didinium nasutum. Ecology 54:1320-1327.

Luckinbill, L.S. 1974. The effects of space and enrichment on a predator-prey system. Ecology 55:1142-1147.

MacArthur, R.H. and E.O. Wilson. 1967. The Theory of Island Biogeography. Princeton Univer. Press, Princeton, N.J. 203 pp.

May, R.M. 1972. Limit cycles in predator-prey communities. Science 177: 900-902.

May, R.M. 1974. Ecosystem patterns in randomly fluctuating environments. Pages 1-50 in R. Rosen and F. Snell (Eds.) Progress in Theoretical Biology. Academic Press, New York.

Maynard Smith, J. 1974. Models in Ecology. Cambridge Univ. Press, Cambridge, England. 146 + xii pp.

Meinhardt, H. and A. Gierer. 1974. Applications of a theory of biological pattern formation based on lateral inhibition. J. Cell Sci. 15:1-27.

Montroll, E.W. 1968. Lectures on nonlinear rate equations, especially those with quadratic nonlinearities. Pages 531-573 in A.O. Barut and W.E. Britton (Eds.), Lectures in Theoretical Physics. Gordon and Breach, New York.

Montroll, E.W. 1972. Nonlinear rate processes, especially those involving competitive processes. Pages 69-91 in S.A. Rice, K.F. Freed, and J.C. Light (eds.) Statistical Mechanics: New Concepts, New Problems, New Applications. Univ. of Chicago Press, Chicago.

Murray,J.D. 1975. Non-existence of wave solutions for the class of reaction-diffusion equations given by the Volterra interacting-population equations with diffusion. J. Theor. Bio. 52:459-469.

Noble, J.V. 1974. Geographical and temporal development of plagues. Nature 250:726-729.

Oaten A. and W.W. Murdoch, 1975. Functional response and stability in predator-prey systems. Amer. Natur. 109:289-298.

Okubo, A. 1972. A note on small organism diffusion around an attractive center: a mathematical model. J. Oceanogr. Soc. Japan 28:1-7.

Okubo, A. 1974. Diffusion-induced instability in model ecosystems: another possible explanation of patchiness. Tech. Rept. 86, Chesapeake Bay Inst., Johns Hopkins Univ.

Okubo, A. 1975. Ecology and Diffusion. Tsukiji Shokan, Tokyo 220 + vii pp.

Oster, G. 1975. Internal variables in population dynamics. This volume.

Oster, G. and Y. Takahashi. 1974. Models for age-specific interactions in a periodic environment. Ecological Monographs 44:483-501.

Othmer, H.G. and L.E. Scriven. 1971. Instability and dynamic pattern in cellular networks. J. Theor. Biol. 32:507-537.

Othmer, H.G. and L.E. Scriven. 1974. Nonlinear aspects of dynamic pattern in cellular networks. J. Theor. Biol. 43:83-112.

Park, T. 1962. Beetles, competition and populations. Science 138:1369-1375.

Patten, B.C. 1968. Mathematical models of plankton production. Int. Revue ges. Hydrobiol. 53:357-408.

Pimentel, D. 1961. Animal population regulation by the genetic feedback mechanism. Amer. Natur. 95:65-79.

Pimentel, D. 1968. Animal population regulation and genetic feedback. Science 159:1432-1437.

Pimentel, D., S.A. Levin, and A.B. Soans. 1975. On the evolution of energy balance in some exploiter-victim systems. Ecology 56:381-390.

Poore, M.E. 1964. Integration in the plant community. J. Ecol. 52 (Suppl.): 213-226.

Reddingius, J. 1971. Gambling for existence: a discussion of some theoretical problems in animal population ecology. Acta Theoretica Supplementum Climum, added to Acta Biotheoretica 20. E.J. Brill and Leiden. 208 + vi pp.

Reddingius, J. and P.J. den Boer 1970. Simulation experiments illustrations stabilization of animal numbers by spreading of risk. Oecologia 5:240-284.

Riley, G.A. 1965. Mathematical model of regional variations in plankton.
    Limnol. Oceanogr. 10 (Suppl.):R202-R215.

Roff, D. 1974. Spatial heterogeneity and the persistence of populations.
    Oecologia 15:245-258.

Root, R.B. 1973. Organization of a plant-arthropod association in simple
    and diverse habitats: the fauna of collards (Brassica oleracea). Ecol.
    Monogr. 43:95-124.

Rosen, G. 1974. Global theorems for species distributions governed by reac-
    tion-diffusion equations. J. Chem. Phys. 61:3676-3679.

Rotenberg, M. 1972.   Theory of population transport, J. Theor. Biol.
    37:291-305.

Rubinow, S.I. 1973. Mathematical Problems in the Biological Sciences, SIAM,
    Philadelphia. 90 + vi pp.

Russell, B.C., F.H. Talbot, and S. Domm. 1974. Patterns of colonization of
    artificial reefs by coral reef fishes. Pages 207-215 in Proc. Second
    Internat. Coral Reef Symp. 1. Great Barrier Reef Committee, Brisbane,
    October 1974

Segel, L.A. and J. Jackson. 1972. Dissipative structure: an explanation and
    an ecological example. J. Theor. Biol. 37:545-559.

Segel, L.A. and S.A. Levin. Application of nonlinear stability theory to the
    study of the effects of diffusion on predator-prey interactions. Proc. AIP
    Conf. (Festschrift for Julius Jackson) (in press).

Simberloff, D.S. and E.O. Wilson. 1969. Experimental zoogeography of islands:
    the colonization of empty islands. Ecology 50:278-296.

Simberloff, D.S. and E.O. Wilson. 1970. Experimental zoogeography of islands:
    a two-year record of colonization. Ecology 51:934-937.

Skellam, J.G. 1951. Random dispersal in theoretical populations. Biometrika
    38:196-218.

Skellam, J.G. 1973. The formulation and interpretation of mathematical models
    of diffusionary processes in population biology. Pages 63-85 in M.S.
    Bartlett and R.W. Hiorns (eds.), The Mathematical Theory of the Dynamics
    of Biological Populations. Academic Press, London.

Slatkin, M. 1974. Competition and regional coexistence. Ecology 55:128-134.

Smith, F.E. 1972. Spatial heterogeneity, stability, and diversity in eco-
    systems. Pages 309-335 in E.S. Deevey (ed.). Growth by in intussusception:
    ecological essays in honor of G. Evelyn Hutchinson. Trans. Conn. Acad.
    Arts Sci. 44.

Steele, John. 1973. Patchiness. Coastal Upwelling and Ecosystems Analysis
    Newsletter 2:3-7.

Steele, J.H. 1974a. Spatial heterogeneity and population stability.   Nature
    248:83.

Steele, J.H. 1974b. Stability of plankton ecosystems. Pages 179-191 in M.B.
    Usher and M.H. Williamson (eds.), Ecological Stability. Chapman and Hall,
    London.

Strathmann, R. 1974. The spread of sibling larvae of sedentary marine inverte-
    brates. Amer. Natur. 108:29-44.

Sutherland, J.P. 1974. Multiple stable points in natural communities. Amer.
    Natur. 108:859-873.

Taylor, D.L. 1973. Ecology. 54:1394-1396.

Thom, R. 1969. Topological models in biology. Topology 8:313-335.

Tikhonov, A.N. 1948. On the dependence of the solutions of differential equa-
    tions on a small parameter. Mat. Sb. 22:193-204.

Tikhonov, A.N. 1952. Systems of differential equations with small parameters
    at the derivatives. Mat. Sb. 31:575-586.

Turing, A. 1952. The chemical basis of morphogenesis. Phil. Trans. Roy. Soc.
    B237:37-72.

Udovic, J.D. and S.A. Levin. 1975. Density dependent selection and genetic
    feedback. M.S.

Vandermeer, J.H. 1973. On the regional stabilization of locally unstable
    predator-prey relationships. J. Theor. Biol. 41:161-170.

von Foerster, H. 1959. Some remarks on changing populations. Pages 382-407
    in F. Stohlman, Jr. (ed.) The Kinetics of Cellular Proliferation. Stratton,
    New York.

Whittaker, R.H. 1953. A consideration of the climax theory: the climax as a
    population and a pattern. Ecol. Monogr. 23:41-78.

Whittaker, R.H. 1974. Climax concepts and recognition. Pages 139-356 in R.
    Knapp (ed.), Handbook of Vegetation Science. Part VIII. Vegetation
    Dynamics. Dr. W. Junk B.V., The Hague

Whittaker, R.H. 1975. The design and stability of plant communities. In 1st
    Internat. Congr. Ecol., The Hague, September 11, 1974.

Whittaker, R.H., S.A. Levin and R.B. Root. 1973. Niche, habitat, and ecotope
    Amer. Natur. 107:321-338.

Whittaker, R.H., S.A. Levin, and R.B. Root. 1975. On the reasons for distin-
    guishing 'niche, habitat and ecotope'. Amer. Natur. 109:479-482.

Wright, H.E. Jr. 1974. Landscape development, forest fires, and wilderness
    management. Science 186:487-495.

Lectures on Mathematics in the Life Sciences
Volume 8, 1976

INTERNAL VARIABLES IN POPULATION DYNAMICS

G. Oster [1]

Contents

1.   INTRODUCTION

Criticizing mathematical models in ecology is like harpooning a blimp: it is almost impossible to miss, and every thrust is likely to be fatal! Most population biologists cut their teeth on models which presume to predict the future course of a population's growth from a knowledge of its present census. That is, models of the form

$$\dot{\underset{\sim}{N}} = \underset{\sim}{F}(\underset{\sim}{N})$$

(1-1a)

or

$$\underset{\sim}{N}_{t+1} = \underset{\sim}{F}(\underset{\sim}{N}_t)$$

(1-1b)

where $\underset{\sim}{N}(t)$ or $\underset{\sim}{N}_t$ is a vector of population numbers $(N_1(t), N_2(t), \ldots N_R(t))$. Criticism of such models has become so ubiquitous in recent years that there is little point to flogging so dormant a horse. It is sufficient to note that population numbers alone do not

---

[1] Work supported by N.S.F. Grant No. BMS74-21240

suffice to describe the dynamics of even that simplest of ecosystems, the bacterial chemostat. It is well known, for example, that the numbers of organisms in a chemostat can
oscillate whilst the biomass remains constant. Clearly numbers and mass must be regarded
as independent state variables (Williams, 1971; Frederickson, 1971).

Our discussion here will be an elaboration on the theme that the demographic trajectory of a population can be decisively affected by its "internal structure". By "internal
structure" we simply mean the totality of coordinates -- other than total population numbers -- which must be specified in order to project the future course of population
growth. Thus, in order to predict the population trajectory we must write an equation of
motion for the density function $n(t, a, \underset{\sim}{x})$ specifying the number of individuals at time t
of age a, and $\underset{\sim}{x} = (x_1, x_2, ..., x_i)$, where $\underset{\sim}{x}$ is a complete set of those characteristics
which influence the population growth rate such as mass, temperature tolerance, aggres-
siveness, and so forth.

Of course, when we attempt to do this we inevitably encounter formidable mathematical
difficulties, and we must ask ourselves whether the investment in effort is worth the return. That is, are there any qualitatively different and dramatically important phenomena
that arise as a consequence of including these internal variables in our models? Or can
we predict the major features of interest to population biologists from models which account for only a single state variable such as total population number or biomass and ignore the internal structure? Just as we would have to look inside a jumping bean to see
why it hops, our discussion here will focus on those circumstances where the internal
structure does affect the dynamics dramatically. In particular, we shall be interested
in the role that age structure plays in shaping the demographic behavior of populations.

2.   The Equation of Motion for the Population Density Function $n(t, a, \underset{\sim}{x})$.

2-1   The equation governing the time evolution of the density function $n(\cdot)$ has
been re-derived by many authors in a number of contexts (e.g. Hulbert and Katz, 1964;
Sinko and Streifer, 1967 ; von Foerster, 1959). The simplest way to arrive at it is by
analogy with the equation of convective flow from fluid mechanics. Let the state of an
individual in the population be specified by a point in $\mathbb{R}^{k+1}$: $p = (a, x_1, ..., x_k)$. Then a
conservation equation for $n(t, a, \underset{\sim}{x})$ can be written straightaway:

$$\frac{\partial n}{\partial t} + \nabla \cdot \underset{\sim}{J} = S \qquad (2-1)$$

where $\underset{\sim}{J}$ is the flux of n in $\mathbb{R}^{k+1}$ and S is the "source and sink" for n , i.e. the
loss of individuals by death and the gain of individuals from births. The model will be

specified when we choose the functional forms for $\underset{\sim}{J}$ and $\underset{\sim}{S}$.

For the flux of individuals through $\mathbb{R}^{R+l}$ we can write

$$\underset{\sim}{J} = \underset{\sim}{V}n \tag{2-2}$$

where $\underset{\sim}{V}$ is the growthrate (or convective velocity)

$$\underset{\sim}{V} = (\frac{da}{dt}, \frac{d\chi_1}{dt}, \ldots, \frac{d\chi_R}{dt}) \tag{2-3}$$
$$= (1, g_1, \ldots, g_R)$$

The growthrate in age is, of course, unity, while the growthrates g_i in trait χ_i may be functions of $t, a, \underset{\sim}{\chi}$ and functionals of n (e.g. $N = \int n \, da \, d\underset{\sim}{\chi}$).

The function S can be decomposed into birth and death contributions: S = birthrate-deathrate. The latter we shall write as

$$\textbf{deathrate} = \mu(\cdot)n \tag{2-4}$$

where μ is the per capita deathrate which can be a function of $t, a, \underset{\sim}{\chi} \, \& \, n$.

The birthrate contribution to S is the feature which gives the population equations their special character. We can model the birthrate phenomenologically in the following way.

$$\textbf{birthrate} = \iiiint da' da'' \, d\underset{\sim}{\chi}' d\underset{\sim}{\chi}'' \, B(t, a, \underset{\sim}{\chi} \mid a', a'', \underset{\sim}{\chi}', \underset{\sim}{\chi}'') \cdot$$
$$n(t, a', \underset{\sim}{\chi}') n(t, a'', \underset{\sim}{\chi}'') \tag{2-5}$$

The kernel function $B(\cdot)$ describes the per capita birthrate, that is the number of newborn with phenotype array $\underset{\sim}{\chi}$ which arise from parents of ages a', a'' and phenotypes $\underset{\sim}{\chi}', \underset{\sim}{\chi}''$. When integrated over all parental types we obtain the total birthrate, $n(t, o, \underset{\sim}{\chi})$. We can include equation (2-4) on the right side of the equation of motion (2-1) by multiplying by a delta function, $\delta(a)$; however, it will prove more convenient to carry the birthrate along as a separate boundary condition.

It is straightforward to write a set of coupled equations to describe interacting populations. However, rather than attempt a general formulation we shall treat some special cases later. We also mention that we have neglected the effect of the spatial distribution of the population. This is a crucial aspect of real populations; however, we shall be dealing mostly with laboratory populations wherein the spatial dimensions of the system are small enough so that the population can be considered spatially homogeneous. A formulation of the migration terms can be found in Rotenberg, 1973. Furthermore, we have neglected dispersion in growthrates by our choice of the flux term (2-2). To account for this we could modify the flux equation to include a "diffusion-like" term:

$\underset{\sim}{J} = \underset{\sim}{\gamma}n + \nabla(\sigma^2 n)$ (Weiss, 1968; Rotenberg, 1976). For the experimental systems we shall be dealing with this effect will generally be negligible.

3. The Role of Internal Structure in the Dynamics of Single Populations: Nicholson's Experiments.

3-1   The Austrialian entomologist A. J. Nicholson performed a meticulous set of experiments on the population dynamics of sheep blowflies; the results of one of these experiments is shown in figure 1.

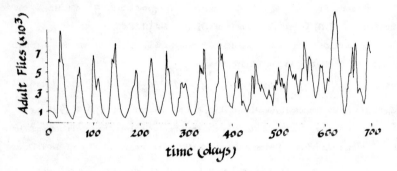

This figure graphs a census of the adult flies over a period of some 700 days, or more than 25 generations under conditions of constant food supply. In particular, he restricted the supply of protein available to the adults while permitting them adequate supplies of sugar and water. The adult flies could live perfectly well on the sugar-water diet, but the females require protein in order to form and lay eggs. When the density of adults is high, Nicholson reasoned, competition for a constant food supply should result in few eggs being laid. Consequently, the following generation would be small ensuring each adult sufficient protein to achieve maximum fecundity. Thus a perpetual sequence of alternating large and small generations should result; figure 1 shows that, indeed, the adult population exhibits a sustained oscillation whose frequency is about 38 days.

In attempting to model this situation we must account for at least two aspects of the system which are central to the existence of the oscillations. First there is the age structure; that is the time delay between the egg-laying by the adults of one generation and the emergence of the next generation's adult population. Second, the dependence of each adult's fecundity on its nutritional history. Thus we must write an equation of motion for the density function $n(t,a,x)$, where x is a measure of the nutritional state, i.e. protein content:

$$\frac{\partial n}{\partial t} + \frac{\partial n}{\partial a} + \frac{\partial}{\partial x}(gn) = -\mu n \qquad (3\text{-}1)$$

The deathrate in Nicholson's experiments was negligible for larvae and pupae, and was nearly random for adults and so we shall assume that $\mu(\cdot)$ is approximately $\mu H(a-\alpha)$ where μ is constant and $H(a-\alpha)$ is a step function at α, the age of adult emergence.

The birthrate we shall approximate by

$$n(t,0,x_0) = \int_0^\infty \int_\alpha^\infty da\,dx'\,n(t,a,x')b(a,x',x_0) \qquad (3\text{-}2)$$

where we shall ignore hereditary effects for the time being and assume that each egg begins its life at the same nutritional state, x_0.

To complete the model we must specify the submodels for the growth and birth functions, $g(\cdot)$ and $b(\cdot)$.

The dependence of egg-laying on protein nutritional state behaves qualitatively as shown in figure 2a; the important feature is that there is a "threshold" level below which few, if any, eggs are laid.

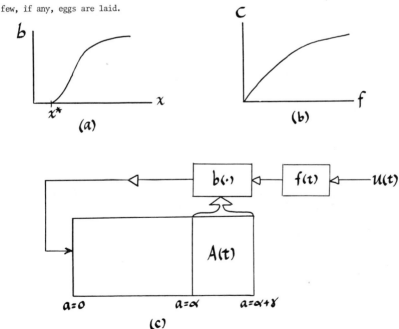

(a)

(b)

(c)

The growthrate in x is governed by the balance between protein consumption and metabolism:

$$g(\cdot) = \frac{dx}{dt} = \text{consumption} - \text{metabolism}$$
$$= C - M \qquad (3\text{-}3)$$

The simplest assumption concerning metabolism is that it follows a first-order rate law:

$$M = kx \qquad (3\text{-}4)$$

(A more reasonable assumption might be a Michaelis-Menten form; however the experiments under consideration here deal with the situation where the food supply is limited, so we would be justified in linearizing a nonlinear relation about x = 0).

The balance equation for the food consumed by the adult flies, f(t), is simply

$$\frac{df}{dt} = \text{supply} - \text{consumption}$$
$$= u(t) - C \cdot N(t) \qquad (3\text{-}5)$$

(Nicholson also conducted a series of experiments to study the "frequency response" of the system by varying the food supply in a periodic fashion. However, we shall deal only with the case $u(t)$ = constant here.)

The consumption term in equations (3-4) and (3-5) is also a potentially complex sub-modelling problem, since competition for a limited protein supply dominates the per capita consumption. A saturating curve similar to that shown in figure 2b is to be expected, although at low supply a linearization is probably appropriate. Figure 2c shows the structure of the model.

Figure 3 shows a simulation of the model equations; all the system parameters were

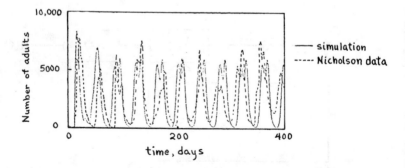

taken from Nicholson's experiments. (Details of the simulation can be found in Oster, Auslander and Allen, 1975). It is clear that the model captures the central features of the dynamics: the amplitude and frequency of the basic limit cycle. There are a number of details, or "fine structure", in the data that require further comment and we shall return to this later.

3-2   While the fact that we can quantitatively predict the population oscillations re-assures us that we have included in the model the relevant biological features, never-theless numerical simulation alone leaves us somewhat dissatisfied. We would like to augment our numerical results with some insight into the mathematical reasons for the oscillations. (The biological reasons for the oscillations were known -- or guessed -- a priori, and built into the model!)

The mathematical mechanism can be inferred from the following fact. By far the most sensitive parameter in the model is the threshold level of protein, χ^*, required to initiate egg-laying (cf. figure 2a). When $\chi^* = 0$, so that there is a continuous increase in fecundity with protein, the system does not exhibit oscillations. As χ^* is increased the system remains quiescent until a critical value is reached whereupon oscillations are initiated. The generation of stable limit cycles in a dynamical system as a parameter is varied suggests we examine the model from the viewpoint of bifurcation theory. The rele-vant theorem is the following, whose proof can be found in Marsden and McCracken (1975).

The Hopf Bifurcation Theorem: Given a vector field $\underset{\sim}{X}(\underset{\sim}{\chi}, \pi)$ on $\mathbb{R}^2$ depending on a parameter, π we view the flow on $\mathbb{R}^3$ as shown in figure 4a. Let the following conditions hold:

(a) $\underset{\sim}{X}(\underset{\sim}{0}, \pi) = \underset{\sim}{0}$ i.e. the $\underset{\sim}{x} = 0$ axis is a fixed point for $\underset{\sim}{X}(\cdot, \pi)$ and the origin ($\underset{\sim}{0}, 0$) is an attractor for $\underset{\sim}{X}(\cdot, \cdot)$.

(b) The Jacobian of $\underset{\sim}{X}$ at the origin, $D\underset{\sim}{X}_\pi(\underset{\sim}{0})$, has a conjugate pair of simple eigenvalues: $\lambda_1(\pi) \pm i\,\lambda_2(\pi)$ such that

(i) $\lambda_1(0) = 0, \lambda_2(0) \neq 0$.

(ii) $\frac{\partial \lambda_1}{\partial \pi}(0) \neq 0$, the eigenvalues move smoothly across the imaginary axis as π increases through some critical value, which we choose as 0.

Then, for some $\varepsilon > 0$, there is a one-parameter family of stable periodic orbits for $\underset{\sim}{X}_\pi$ when $0 < \pi < \varepsilon$. The situation is shown in figure 4b.

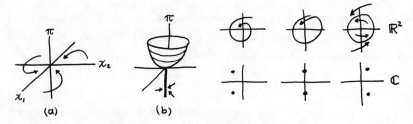

Fig. 4

In other words, if the origin $\underline{x} = \underline{0}$ is a stable attractor for $\pi \leqslant 0$, and if as we tune the parameter π through zero the eigenvalues smoothly cross the imaginary axis away from the real line, then the formerly stable origin will bifurcate to a stable limit cycle, for some range of $\pi > 0$. (If the eigenvalue crosses on the real line, a bifurcation may occur, but it will not be to a periodic orbit.)

Although we have illustrated the theorem in two dimensions the extension to $\mathbb{R}^n$ is straightforward using the Center Manifold Theorem (cf. Marsden and McCracken, op. cit.). The generalization to infinite dimensional systems can be found in Marsden (1974).

The conditions of the theorem are, of course, quite difficult to check for our system; however, a judicious use of numerical simulation suffices when the parameters are fixed by the experimental data. In particular, a numerical root locus of the linearized system locates the bifurcation point and a simulation of the nonlinear system in this vicinity verifies the "vague attractor" conditions.

3-3   The next step in our analysis is to try and reduce the model to a more comprehensible scale by making certain approximations. In this way we can hope to deepen our insight into the functioning of the model while lessening our dependence on numerical simulations.

The first approximation we might try is to eliminate the internal variable χ which bookkeeps the nutritional state of the population. We can do this by making the "quasistatic" approximation $\dot{\chi} = \dot{f} = 0$. That is, we assume that the time scale for the physiological processes of feeding and assimilation is much shorter than the demographic time scale on which the limit cycles occur. From equations (3-4) and (3-5) we can express x as a function of the adult population, $N(t) = \iint n\, da\, d\chi$:

$$\chi = \frac{u(t)}{k\, N(t)} \tag{3-6}$$

Then integrating equation (3-1) over all x, the model reduces to

$$\frac{\partial n}{\partial t} + \frac{\partial n}{\partial a} = -\mu n \tag{3-7a}$$

$$n(t,0) = \int_{\alpha}^{\alpha+\gamma} b(N)n\,da = b[N]N \qquad (3\text{-}7b)$$

$$N(t) = \int_{\alpha}^{\alpha+\gamma} n\,da \qquad (3\text{-}7c)$$

where $b[N]$ is a decreasing function of the adult population, $N(t)$. By removing
the dynamics from the feedback loop in figure 2c we have reduced the model to one involving the adult population only: integrating (3-7a) over all ages and substituting 3-7b
we obtain the function equation in N(t):

$$\frac{d}{dt}N(t) = l(\alpha)\,b[N(t-\alpha)]N(t-\alpha) - \bar{\mu}N(t) \qquad (3\text{-}8)$$

where $l(\alpha) = e^{-\int_0^\alpha \mu\,da}$ is the fraction of newborn which survive to age α .

3-4   In figure 5 is shown a simulation of equation (3-8). Not only is the major limit
cycle still evident, but we see that the characteristic "double peak", evident in both
the experimental data and the original model, is still present. However, from the reduced
model (3-8) it is easy to see where this "superharmonic" component comes from. Notice
that the first term on the right hand side of (3-8) -- the net egg-laying rate times the
survivorship -- has a single critical point, N_c (since $b[N]$ is monotone decreasing).

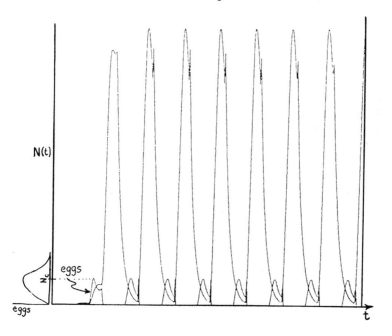

In figure 5 we have drawn the net egg-laying rate to the left of the time plot. We see that if the amplitude of the limit cycle is greater than N_c the egg-laying rate will be bimodal (i.e. the function $Nb[N]$ acts as a "frequency doubler"). Thus, if the pre-adult mortality, $l(\alpha)$ is sufficient the adults too will have a bimodal peak. Moreover, as shown in figure 4c, if a small amount of randomness is added to the parameters $\alpha, \mu \ \& \ b$ the appearance of the simulation of equation (3-8) resembles the data even more closely.[*]

3-5   Remark: The onset of the limit cycle can be calculated by computing the point where the system, linearized about its equilibrium $b[\bar{N}] = \frac{\mu}{l}$, becomes unstable. The linearized system is

$$\dot{x} = c_1 x(t-\alpha) - \bar{\mu} x(t)$$

$$(3-9)$$

where c_1 is the linearization constant, $\frac{d}{dN}(lb[N]N)\big|_{\bar{N}} = \mu \ \& \ x = N - \bar{N}$. The characteristic equation is

$$s = c_1 e^{-s\alpha} - \mu , \quad s \in \mathbb{C}$$

Choosing an explicit form for $b[N]$ we can compute the conditions for instability by calculating when the single real root of (3-10) becomes positive. For example, if $b[N] = be^{-kN}$, or $b/1+\bar{N}$ the linearized system becomes unstable when $lb = \mu$, as expected. Moreover, the lowest pure imaginary root gives us the linear estimate for the period of the limit cycle: $\alpha < T < 2\alpha$, where α is the age of adult emergence ($\sim$ 18 days). The actual cycle length is approximately 38 days; later a nonlinear analysis will give us a better estimate.

3-6   We can further simplify our model by incorporating one more feature of the blow-flies' reproductive physiology. It turns out that a female does not lay eggs continuously over her entire adult lifespan, but does her ovipositing in "bursts" every few days -- always in the morning -- depending on her nutritional state. Therefore, we can approximate $b(a, N)$ by:

$$b(a, N) = \sum_{i=1}^{K} b_i[N] \delta(a - \alpha_i)$$

$$(3-10)$$

---

[*]In the simulations shown the birthrate $b[N]$ was approximated by the function $bn \exp(-kN)$, where b is the maximum fecundity and k measures the strength of the competition for food.

The $b_i[\cdot]$ have the same density dependence as before. Using (3-10) in equations (3-7) the model reduces to the following coupled set of difference equations:

$$B(t) \triangleq n(t,o) = \sum_i b_i[N] l(\alpha_i) B(t-\alpha_i) \quad (3\text{-}11a)$$
$$N(t) = \int_o^\infty l(a) B(t-a) da \quad (3\text{-}11b)$$

In this form the model is equivalent to a "density-dependent Leslie matrix":

$$\underset{\sim}{n}_{t+1} = \underset{\sim}{L}(n) \underset{\sim}{n}_t \quad (3\text{-}12)$$

where (3-11a) is the equation for the top row of $\underset{\sim}{L}$ and $\underset{\sim}{n}_t \in \mathbb{R}^N$ is a vector of N age classes. We could have obtained (3-13) directly from (3-7) by finite differencing forward in time and backward in age (Oster, 1976).

Figure 6 shows a simulation of this model. In addition to retaining the features characteristic of the previous formulations we observe a new feature: the peaks of the

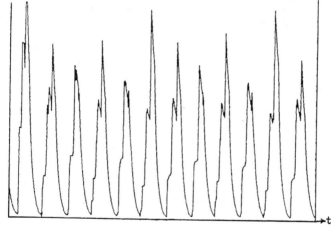

major limit cycle appear to vary in amplitude in a regular fashion. In order to investigate the origin of this "subharmonic" we shall examine the special case of equation (3-11) where i = 1, i.e. only one reproductive burst (or, equivalently, a 1 x 1 Leslie matrix). Thus equations (3-11) reduce to the difference equation on $\mathbb{R}^1$:

$$B(t) = l(\alpha) b[N] B(t-\alpha)$$
$$N(t) = \int_\alpha^\infty l(a) B(t-a) da$$

or, taking $\alpha \triangleq 1$,

$$B_{t+1} = \varphi(B_t) B_t$$
$$\triangleq F(B_t) \qquad (3\text{-}13)$$

where $\varphi(B_t) = \varphi\left[\int_a^{\infty} l(a) B(t-a) da\right]$ is a monotone decreasing function of B_t. In figure 7 is shown a graph of $F(\cdot): \mathbb{R}' \rightarrow \mathbb{R}'$.

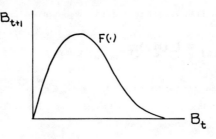

Fig. 7

The function graphed in figure 7 is prototypical of a large class of difference equations that have been employed in the entomological literature to model insect populations with nonoverlapping generations. Equation (3-13), therefore, is interesting in its own right, aside from its relevance as an approximation to our original model. Therefore, we shall briefly digress in order to summarize some properties of nonlinear difference equations. A more complete discussion can be found in Guckenheimer, Oster and Ipaktchi (1976) (hereafter referred to as GOI) and May and Oster (1975).

4.   Nonlinear Difference Equations

4.1   In order to illustrate the general theory we shall focus our attention on a particular example. Consider the map $f: \mathbb{R}' \rightarrow \mathbb{R}'$ defined by

$$f(x) = rx(1-x) \, , \quad 0 < r \leq 4 \qquad (4\text{-}1)$$

This function is graphed in figure 8a for several values of the parameter, r. Here, r may be interpreted as the net reproductive rate, and we have restricted its range so as to ensure the population level is always positive.[*]

---

[*] A better model for equation (3-13) might be
$$g(x) = rx \exp[a(1-x)] \qquad (4\text{-}2a)$$
or
$$h(x) = rx/(1+ax^b) \qquad (4\text{-}2b)$$
However, the qualitative features we wish to discuss are the same for all three models (cf. May and Oster, 1975; Guckenheimer, Oster and Ipaktchi, 1976).

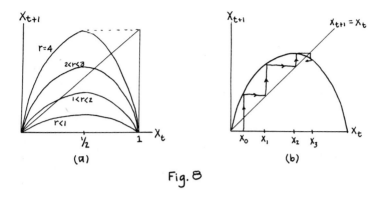

Fig. 8

For $r > 1$, $f(\cdot)$ has one nontrivial fixed point $f(\bar{x}) = \bar{x}$ which, for $r > 2$ lies to the right of the critical point, $x_c = \frac{1}{2}$. As figure 8b shows, the approach to equilibrium can be projected out graphically; for $1 < r < 2$ the trajectory settles monotonically to $\bar{x}$ while for $2 < r < 3$ the approach is oscillatory. In this section we shall study the behavior of equation (4-1) as the parameter is varied between 3 and 4.

4-2   The local stability of the fixed point $\bar{x}$ is determined by the eigenvalue of the map linearized about the fixed point:

$$\lambda(r) = f'(\bar{x}) = 2 - r \qquad (4-3)$$

The fixed point is attracting so long as $|\lambda(r)| < 1$ $i.e.\ r < 2$. When r is increased past 2, however, $\bar{x}$ becomes repelling and the orbit winds out from $\bar{x}$. To see what happens next we examine the second iterate of $f(\cdot)$: $f^{(2)} \equiv f \circ f(\cdot)$. The graph of $f^{(2)}$ is shown in figure 9a for $r = 3$. We see that as r increases past 3 $f^{(2)}$ intersects the 45° line in three points. At $r = 3$ the fixed point at $\bar{x} = \frac{2}{3}$ bifurcates into two new fixed points of period 2: $f^{(2)}(\bar{x}_i^{(2)}) = \bar{x}_i^{(2)}, i=1,2.$ The period-2 orbit remains stable so long as the eigenvalues of $f^{(2)}(\bar{x}_i^{(2)})$, $\lambda_i^{(2)}(r) = f^{(2)\prime}(\bar{x}_i^{(2)})$ lie within the unit circle $|\lambda_i^{(2)}(r)| < 1$. By the chain rule, $\lambda_i^{(2)}(r) = f'(\bar{x}_1^{(2)}) \cdot f'(\bar{x}_2^{(2)})$, so the criteria for the stability of the 2-point cycle is $\lambda^{(2)}(r) < 1$: the product of the slopes of $f^{(2)}$ at $\bar{x}_1^{(2)} \& \bar{x}_2^{(2)}$. As the parameter is increased further the period-2 orbit becomes unstable when $\lambda_i^{(2)}(r) < -1$ and a stable 4-point cycle appears whose stability is governed by $\lambda_i^{(4)}(r) = \prod_{i=1}^{4} f'(\bar{x}_i^{(4)})$. By this mechanism cycles of length 2^k, $k = 1, 2, \ldots$ are generated as the parameter is increased. However, the range of parameter values for

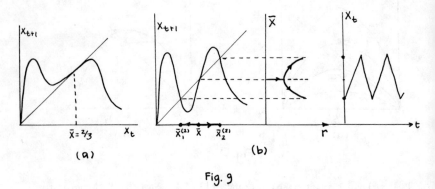

Fig. 9

which the 2^k-point cycle is stable grows smaller as k increases. Eventually, a critical value of $r = r^*$ is reached where a second kind of bifurcation occurs. This happens when one of the minima of $f^{(k)}$, for some k, dips low enough to touch the 45° line. When this happens a bifurcation through $\lambda(r) = +1$ occurs which creates a new fixed point at the point of tangency as shown in figure 10. This new fixed point promptly splits into two

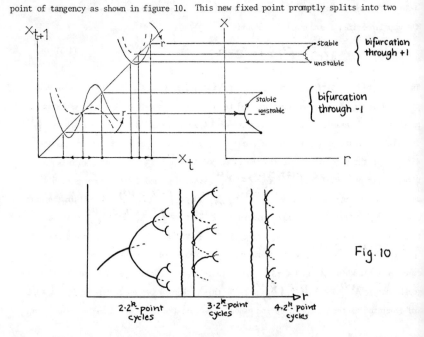

Fig. 10

fixed points, one stable and one unstable. Li and Yorke (1975) have proven that when a period-3 cycle appears then there exist cycles of any length as well as orbits which are aperiodic, i.e. which asymptotically approach neither an equilibrium point nor a periodic orbit. For the 1-dimensional maps (4-1), (4-2a) and (4-2b) two facts can be proven: (GOI): (i) there is at most one stable periodic orbit for each value of the parameter, (ii) the aperiodic points have nonzero measure only at r = 4.

Using a technique from dynamical systems theory called "symbolic dynamics" it is possible to compute the number of coexisting periodic orbits for any parameter value (GOI). For equation (4-1) it turns out that the number of fixed points, N_k, of period k form a Fibonacci sequence: $N_k = \left\{ 1, 3, 4, 7, 11, \ldots \right\}$.

4.3   When r = 4, then f has no attracting periodic points (despite the fact that the periodic points are still dense in the interval $[0, 1]$). The orbit of $\hat{f} = 4x(1-x)$ does not asymptotically approach either an equilibrium or a periodic orbit, but rather approaches a much more complicated limit set called a "strange attractor" (Ruelle and Takens, 1971). Formally, a strange attractor for a map f is defined to be an infinite set Λ with the following properties: (i) $f(\Lambda) = \Lambda$, i.e. Λ is invariant under f; (ii) f has an orbit which is dense in Λ ; (iii) Λ has a neighborhood $\mathcal{a}$ consisting of points whose orbits tend asymptotically to Λ : $\lim_{t \to \infty} f^{(t)}(a) \subset \Lambda$.

While strange attractors are not structurally stable for 1-dimensional maps, they are quite robust in higher dimensions. Their most important property is that orbits in or near such a limit set behave in an essentially chaotic fashion; consequently the dynamical description of orbits must be couched in statistical terms. For example, we can give a statistical description of the map $\hat{f}(x) = 4x(1-x)$ by computing the density function giving the distribution of points on $[0, 1]$ after a large number of iterations (Stein and Ulam, 1964):

$$\rho(x) = \frac{1}{\pi \sqrt{x(1-x)}} .$$

It can be shown that $\hat{f}$ is topologically equivalent to the piecewise linear map $\tilde{f}(x) = 2x \ (0 \leqslant x \leqslant \frac{1}{2}), = -2x \ (\frac{1}{2} \leqslant x \leqslant 2)$ shown in figure 11 by the homeomorphism $h(x) = \frac{2}{\pi} \sin^{-1} \sqrt{x}$. If we agree to count points in the left half of the unit interval as "0" and points in the right half as "1", then the orbit of $\tilde{f}$ generates a sequence of 0's and 1's whose density function is uniform. That is, the orbit of $\tilde{f}$ is equivalent to a sequence of Bernoulli trials (e.g. flips of a fair coin). This example graphically illustrates the fact that the deterministic dynamics in a strange attractor can be, literally, as random as possible.

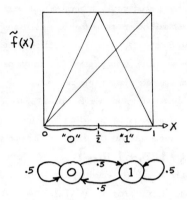

4.4   The lesson to be drawn from this discussion is the richness of the dynamical behavior of which so simple an equation is capable: equilibria, cycles of any length as well
as totally chaotic behavior. The situation, as we shall see, is even worse in higher
dimensions. What does this analysis tell us about the behavior of our model as it relates
to Nicholson's data?

First, we see that the initial bifurcation creates a 2-point cycle whose length is
perforce 2α : twice the age to maturity (recall our time step is in units of α). This
substantiates our linear analysis in section 3-5 and is in accord with the observed limit
cycle period.

Second, an examination of the data reveals a repeating sequence of three peak heights
over the first 400 days (i.e. the first peak is the highest, thence a rising sequence of
three, which repeats). A Fourier spectrum does indeed show a component with a period of
about three cycles. Thus we are led to suspect that the system's periodic behavior is a
reflection of a higher order bifurcation than the first. It is important to note several
things in drawing this conclusion: (i) the difference equation (3-12) has many age classes
while our analysis is based on equation (3-13) which has but one age class. The higher
dimensional system is capable of a much richer spectrum of periodic orbits. For example,
comparing figure 1 with the 1-dimensional model one might suspect that the system is in
an 8-point cycle (after three bifurcations). However, one can verify that the sequence of
peak heights for the stable 8-point cycle is (numbering the branches from highest amplitude to lowest): 1-8-4-5-2-7-3-6. (It is easy to verify the pattern: "the higher the peak,
the lower the crash"). This amplitude sequence is not the same as the 8-point cycle in

the data. Equation (4-19), however, can produce such a cycle under multiple bifurcations. (ii) As each successive bifurcation occurs a new component appears in the frequency spectrum. However, the old component does not disappear, but attenuates gradually. For example, the 4-point cycle shows strong traces of the 2-point cycle (since the orbit is forced to alternate between the upper branch pair and the lowest). Thus while the data is dominated by the basic cycle one should expect to see lower frequencies superimposed indicating multiple bifurcations.

If the multiple bifurction explanation of the major limit cycle and the superimposed subharmonics be true then we have in hand also a ready explanation for the apparent descent into chaos of the data after the 400th day. Recall that the domains of stability along the parameter axis of the cycles decreased as the cycle lengths increased. Thus, for higher order cycles small changes of the parameter can force the system through a large number of higher branches initiating, quite rapidly, cycles of much greater length, or even chaotic orbits. Moreover, upon closer examination the data between 400 and 700 days appears to retain many of the periodic features of the preceeding cycles and is probably not yet into the chaotic regime.

From a practical standpoint it is probably irrelevant whether an orbit is aperiodic or is in a long stable cycle. Seldom will there be a data record of sufficient length or resolution to discriminate the two. This raises some unsettling issues with respect to the usefulness of deterministic models in population ecology, a subject discussed more fully in May and Oster (1975).

4.5   Next, let us add one more age class to the difference equation (3-12), and study the 2-dimensional Leslie model:

$$F : \mathbb{R}^2 \longrightarrow \mathbb{R}^2$$
$$(x,y) \longmapsto F(x,y) = \left[\beta(x,y), \; S(x,y)x \right] \quad (4-3)$$

where $\beta(x,y)$ is the density-dependent birthrate and $S(\cdot)$ is the fraction of individuals in the first age class which survive to the second age class. Once again, for purposes of exposition, we shall adopt particular functional forms for $\beta(\cdot) \; \& \; S(\cdot)$. In particular, for $\beta(\cdot)$ we shall employ the exponential density-dependent factor (2-2a):

$$\beta(x,y) = (b_1 x + b_2 y) e^{-a(x+y)} \quad (4-4a)$$

The coefficient b_1 and b_2 give the maximum per capita birthrates for each age class. In GOI the following special case is thoroughly investigated: $b_1 = b_2 = r$, $s=1$ and $a=0.1$, i.e.:

$$F(x,y) = \left[r(x+y) e^{-0.1(x+y)}, \; x \right] \quad (4-5)$$

Equation (4-3) depends on the single parameter, r. Figure 9 shows the complete bifurcation

structure for equation (4-5) as r is varied. It differs in several important respects
from the 1-dimensional case, and so we shall briefly describe its major features.

From equation (4-5) the equilibrium population is

$$\bar{x} = \bar{y} = \frac{\ln 2r}{0.2} \qquad (4\text{-}6)$$

The local stability of this equilibrium is determined by linearizing about $(\bar{x}, \bar{y})$:

$$DF(\bar{x}, \bar{y}) = \begin{bmatrix} B & B \\ 1 & 0 \end{bmatrix} \qquad (4\text{-}7)$$

where $B \equiv r(1 - 0.1(\bar{x} + \bar{y}) e^{-0.1(\bar{x} + \bar{y})}$

The characteristic polynomial is

$$\lambda(B) = \lambda^2 - B\lambda - B \qquad (4\text{-}8)$$

As r is increased the roots of (4-8) move toward the unit circle and cross when
$r = \frac{e^3}{2} \cong 10.04$. At the point of crossing the characteristic roots of (4-8) are given by

$$\lambda^* = -\frac{1}{2} \pm i\frac{\sqrt{3}}{2} \qquad (4\text{-}9)$$

We recognize this as the two complex cube roots of unity, $z^3 = 1$. Thus near $(\bar{x}, \bar{y})$,
DF looks like a rotation by $\frac{2\pi}{3}$. In figure 12 is shown the 3-point cycle which
appears at $r = \frac{e^3}{2}$. The mechanism which gives rise to this periodic orbit is governed by

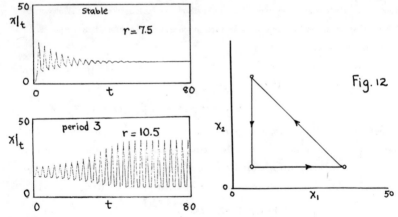

Fig. 12

the Hopf bifurcation theorem for maps (Ruelle and Takens, 1971; Marsden and McCracken,
1976), which is analogous to -- but more complicated than -- the bifurcation theorem for
vector fields we cited earlier. This kind of bifurcation has no analog in the 1-dimen-
sional case discussed above where the eigenvalues are constrained to the real line.

A more detailed examination of the model reveals that the 3-point cycle existed even before the bifurcation at r=10.04 (GOI). The situation is illustrated schematically in

figure 13. Below r=8.95 the equilibrium (4-6) is globally attracting. At 8.95 a

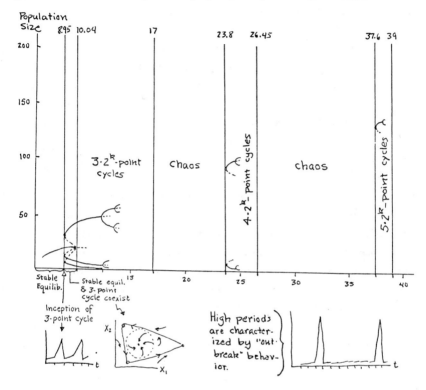

Fig. 13

bifurcation through z=+1 creates a 3-point cycle whose stability is saddle-like. Above

8.95 this promptly splits into a stable and an unstable cycle. As r is increased toward

$e^3/2$ the amplitude of the stable cycle grows while the unstable cycle shrinks. At $e^3/2$

the unstable orbit coalesces with the stable equilibrium; this is the Hopf bifurcation

discussed above. Thus for $8.95 < r < 10.04$ the system can exist in either of two attractors:

large amplitude population vectors are drawn into the 3-point cycle whereas populations

near $\ln(2r)^5$ settle to a stable equilibrium.

As r is increased past 10.04 a sequence of bifurcations through z=-1 takes place

which double the period, producing cycles of length $3 \cdot 2^k$, $k=0,1,\ldots$ similar to the

1-dimensional case. Eventually, near r=17 the system enters a region of chaos. Unlike the 1-dimensional case this region persists for a finite parameter interval $17 < r < 23.8$. Within this interval the orbit is drawn to a true strange attractor wherein the only way to characterize the system behavior is by statistical means.

Indeed, using symbolic dynamics a "statistical mechanics" for population models in the chaotic regime can be developed (GOI). Roughly, it depends on partitioning the strange attractor appropriately and then determining a matrix of transition probabilities between the different parts of the attractor. Then classical Markov chain theory can be employed to characterize the orbit within the attractor. A simple example of this procedure was the partitioning of the piecewise linear map discussed in section 4-2 (cf. figure 11). In that case the transition matrix turned out to be $\underset{\sim}{A} = \begin{bmatrix} .5 & .5 \\ .5 & .5 \end{bmatrix}$ corresponding to a 2-state Bernoulli process.

For the symmetric case ($b_1 = b_2$) under discussion here the trajectory in the strange attractor displays a mixture of regular and chaotic features. At its first appearance the strange attractor nucleates in three pieces about the apexes of the 3-point cycle. As the parameter is increased the time trajectory has the appearance of a 3-point cycle with some random variation in the amplitude of the points of the cycle. This superimposed randomness increases as the parameter is tuned through the range $17 < r < 23.8$. Near the middle of this parameter interval the three segments of the strange attractor merge, and the orbit appears totally stochastic. Toward the end of the interval the attractor begins to condense about a 4-point cycle. At r = 23.8 a bifurcation through z=+1 creates a new fixed point and a period-4 cycle appears. The entire sequence of period-doubling bifurcations through z=-1 repeats, creating cycles of length $4 \cdot 2^k, k=0,1,\cdots$ until, at r=26.45, another strange attractor appears, this time nucleating about the 4-point cycle.

This pattern of alternating regions of chaos and $n \cdot 2^k$-point cycles appears to continue indefinitely, as shown in figure 13. We note, however, that the bands of stable $n \cdot 2^k$-point cycles become narrower as n increases, so high period cycles become more difficult to observe since the bifurcation boundaries become more closely spaced. The reasons for this are discussed in GOI. In practice, as we mentioned earlier, it is probably impossible to distinguish long cycles from chaos.

When the reproductive rate of the two age classes are unequal, $b_1 \neq b_2$, the bifurcation structure is more complex. Instead of a 3-point cycle, the initial bifurcation produces a closed invariant curve traced out by a 3-point cycle which "doesn't quite close". This phenomenon is also anticipated by the Hopf theorem (cf. GOI: Appendix D) and corresponds to a bifurcation through $|z| = 1$ at an irrational angle (i.e. not at one of the low roots of unity $z^k = 1$; thus the Jacobian of $\mathbf{F}$ looks like a rotation by an irrational angle).

4-6   The above discussion gives a general picture of what we should expect in higher dimensions when more age classes are added. Figure 6 shows a simulation of equation (3-12) using 72 age classes. Clearly, so high dimensional a model can support a rich bifurcation structure. A complete study of the dynamical behavior of this model is presented elsewhere (Ipaktchi, et al., 1976); here it suffices to note that as the dimensionality of the model increases the age-specific birthrate levels required to initiate cyclic behavior and chaos descreases sharply. Furthermore, we have noted earlier that the raison d'etre for the limit cycle was the inter-age-class competition for food. However, from the viewpoint of the system nonlinearities generating the bifurcations interspecific competition can accomplish the same thing. Thus, far from being a mathematical curiosity the phenomenon of "deterministic chaos" may be a ubiquitous features of population dynamics in nature. The question of how to distinguish this sort of chaos from "real" stochastic influences remains open.

5.   Two Population Systems

5-1   In two independent experiments Hassell and Huffaker (1969) and Podoler (1974) observed the following remarkable phenomenon on laboratory populations of parasitic wasps preying on moth larvae. Alone, the host population settled into a stable age distribution after 5 or 6 generations. When the parasite was introduced, however, the two populations quickly synchronized into practically discrete generations. These discrete generations consisted of cohort waves propagating through the age structures of each population. The interesting features of these population waves was that they synchronized, out of phase, in such a way that the two populations were able to coexist indefinitely in a rather homogeneous environment, in defiance of some of the conventional wisdom in ecology. In this section we shall indicate how this phenomenon can be understood in terms of bifurcation theory. A more complete account can be found in Auslander, Oster and Huffaker, 1974.

5-2   The central biological feature of this system is the age-specific nature of the interaction between the parasite and host populations. This is shown schematically in figure 14. A model of this situation consists of two conservation equations in age and time:

$$\frac{\partial p}{\partial t} + \frac{\partial p}{\partial a} = -\mu_p p \tag{5-1a}$$

$$\frac{\partial h}{\partial t} + \frac{\partial h}{\partial a} = -\mu_h h \tag{5-1b}$$

$$p(o,t) = \int b_p(a, \tilde{P}, \tilde{H}) p \, da \qquad \text{(5-2a)}$$

$$h(o,t) = \int b_h(a) h \, da \qquad \text{(5-2b)}$$

where $p(t,a)$ and $h(t,a)$ are the parasite and host density functions, respectively $\tilde{P} = \int_{\alpha_p}^{\alpha_p + \gamma_p} p \, da$ and $\tilde{H} = \int_{\alpha_0}^{\alpha_0 + \gamma_0} h \, da$ are the adult parasites and larval hosts, respectively (cf. figure 14).

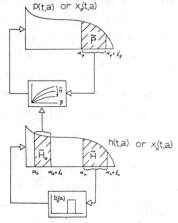

The two populations are coupled by a submodel for the parasitization process. The simplest assumption is a random encounter process which yields a coupling submodel of the form:

$$\text{probability of parasitization} \sim \tilde{H}_o (1 - e^{-A\tilde{P}^k}) \qquad \text{(5-3)}$$

where A is the "area of discovery" and k the "quest constant" (i.e. a correction for density dependence (Hassell and Varley, 1969). Using the parameter values obtained from the Huffaker experiments one can numerically simulate the model and observe the evolution of the age profiles as the parameters A and k governing the interaction strength between the populations are varied. When the interaction is weak (i.e. little parasitization) the parasites die out and the host density settles into a stable age distribution as shown in figure 15a. As the interaction strength is increased the system suddenly bifurcates to a finite amplitude limit cycle. The effect of this on the age profiles of each population is shown in figure 15b and 15c. Here a "stroboscopic snapshot" of the age density functions every generation is plotted. We see that both host and parasite age structures have condensed into travelling waves, with a fixed phase relationship. The

mechanism by which the populations coexist can be seen by plotting the interacting segments of the populations, $\tilde{H}_o \,\&\, \tilde{P}$ (fig. 15d). When the cohort wave in the parasite

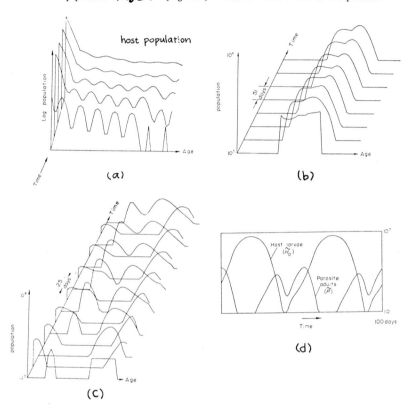

Fig. 15

population reaches the adult life stages the host larval population is at its minimum
value. From figure 15c we see that the "frequency" of the parasite waves is just twice
that of the host. Thus the disynchrony of the generations forces the adult parasite
wave to live off the tails of the host larval waves.

5-3   The mathematical mechanism underlying the limit cycle is the same as that discussed in section 4 for a single population. In this case, however, the bifurcation is
controlled by two parameters (A, k) rather than one, and the limit cycle appears at
finite amplitude rather than growing from zero amplitude (analogous to the bifurcation

at 10.04 in figure 13). The relevant generalization of the Hopf theorem cited earlier
was proven by Takens (1974) for the finite dimensional case. The generic behavior of
the 2-parameter Hopf bifurcation is sketched in figure 16.

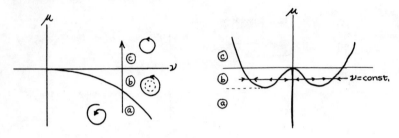

$$Fig.16$$

5-4   The degree of synchronization was so complete in these experiments that Hassell
and Huffaker were able to employ a discrete generation difference equation model of
the form $H_{t+1} = F(H_t, P_t)$, $P_{t+1} = G(H_t, P_t)$ to fit their data. The bifurcation behavior of
host-parasite difference equations of this type have been studied numerically by Bedding-
ton, et al. (1975). Their studies are covered by the theory developed in GOI. In particular, the initial bifurcation from equilibrium produces an invariant circle rather
than a periodic orbit (for reasons discussed in section 4). The strange attractor
then grows from this circle followed by alternating regions of periodic orbits and chaos
analogous to figure 13. Simulation studies of the high dimensional model (5-1)-(5-2) show
that 2-population systems whose interactions are age-specific are even more prone to
chaotic behavior than models without age structure. A more delicate issue -- as yet unresolved -- is the proclivity of the model system to crash if the phase synchrony is up-
set (i.e. one or both populations go to zero). Stated in ecological terms the two populations can create a "temporal" or "phase" niche by shaping their age structure appropriately. However, the resiliency of the populations to perturbations which disrupt this
synchrony is not unlimited. Thus the domain of attraction of the limit cycle is small
with respect to phase disturbances.

6.   Genetic Structure

6-1   Returning to Nicholson's experiments once again we see from figure 1 that the
major limit cycle begins to become disordered after day 400. Nicholson's explanation for

this was that the same intra-specific competition which created the limit cycle simultaneously initiated a process of genetic selection favoring those flies capable of reproducing at low protein levels. Thus, after many generations the average reproductive
rate of the population should rise. Recalling that the bifurcation parameter in the model was the maximum per capita birthrate, b, we see that such a selection process can
initiate a sequence of bifurcations which would push the system closer to the chaotic regime. In fact, Nicholson did determine that the fecundity of flies which survived the
duration of the experiment was higher than a control group. Therefore, we are forced to
consider the genetic structure of the population if we are to model its long-term dynamics. Indeed, this is a central feature of all biological populations. To regard a population as a collection of identical individuals is to ignore their principal characteristic: the ability to adapt to changing environments. Nicholson's experiments are an intriguing illustration of the interaction between genetic and demographic effects. In
this section we shall outline one approach to incorporating genetic variables into our
population model and illustrate how the changing genetic structure can influence the
demographic dynamics by initiating bifurcations.

6-2   The discipline of population genetics is one of the most mathematically highly
developed areas of biology, and the reader can find any number of texts on the subject
(e.g. Jacquard, 1974, Crow and Kimura, 1970). Unfortunately there has been a substantial
gap between population genetics and population ecology. The difficulty is that the
mechanics of Mendelian inheritance are quite complicated when expressed in mathematical
terms. Thus very little is known about the properties of genetic equations except in
special cases and these cases generally bear little relevance to ecological situations.
In particular, traits relevant to complex behavioral and/or physiological performance are
controlled by many genes, while the mathematical apparatus of population genetics is confined to treating one or a few loci, (with one or a few alleles at each locus). Indeed,
the mechanics of polygenic systems is only beginning to be worked out (e.g. L. Hood, et
al., 1975). Therefore, we shall employ a "thermodynamic" formulation which averages out
all of the "statistical mechanical" details of the underlying genetic processes. Such a
"continuum" model has been proposed by Slatkin (1970) and employed by Roughgarden (1972)
and others in a number of ecological contexts. Here we shall briefly describe the
approach; a more complete description can be found in Rocklin and Oster (1975).

For expository purposes we shall concern ourselves with the evolution of the density
function for the case when the generations do not overlap. Therefore, we can ignore the
age distribution and regard time as a discrete variable, so that the equation of motion

62 G. OSTER

can then be written as a difference equation relating $n_t(\xi)$ to $n_{t+1}(\xi)$, where ξ is the phenotypic trait under consideration:

$$n_{t+1}(\xi) = \Psi[n_t(\xi)] \tag{6-1}$$

After we determine the form of the functional $\Psi[\cdot]$ we can easily re-introduce the continuous dependence on age and time.

We shall summarize all of the genetic processes in a "scattering kernel" $L(\cdot)$ as follows.

Let:

$$L(\xi|\xi_1,\xi_2) = \begin{cases} \text{distribution of offspring} \\ \text{phenotypes arising from} \\ \text{matings of parental types} \\ \xi_1 \text{ and } \xi_2 \end{cases} \tag{6-2}$$

(cf. figure 17)

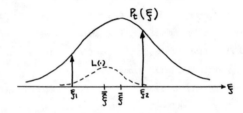

Fig. 17

$L(\cdot)$ is a conditional probability, $\int_0^\infty L\,d\xi = 1$, which contains a phenomenological description of the transmission of quantitative ("metric") traits by sexual reproduction; its functional form must be derived from the mechanics of Mendelian inheritance. Alternatively, it can be measured empirically. For example, many quantitative traits tend to be distributed normally about the parental mean: $\bar{\bar{\xi}} = \frac{1}{2}(\xi_1 + \xi_2)$, so that $L(\cdot)$ can be considered a function of a single variable $\zeta = \xi - \bar{\bar{\xi}}$. Other genetic processes such as heritability and dominance deviations can be included by suitably modifying the form of $L(\cdot)$. (Lande and Slatkin, 1975; Rocklin and Oster, 1975).[*] We emphasize that we are treating the selection process phenomenologically at the phenotypic, not genotypic level. Therefore, we are restricted to discussing genetic traits controlled by many loci with a high degree of heterozygosity. For a treatment of 1-locus, 2-allele theory including

---

[*] Heritability, h^2, measures the tendency of the offspring distribution to regress to the population mean $\bar{\xi} = \int \xi n(\xi)d\xi / \int n(\xi)d\xi$. Thus we can redefine ζ as the convex combination $\zeta = \xi - (h\bar{\bar{\xi}} + (1-h^2)\bar{\xi})$. Dominance deviations tend to make the offspring distribution assymmetric.

age structure see Charlesworth, 1973.

In order to write the equation of motion for $n_t(\xi)$ we must compute the frequency of matings between parental types ξ_1 and ξ_2. If mating is random with respect to the trait then, in a large population, we expect the number of matings to be the product of the parental frequencies: $p_t(\xi_1)p_t(\xi_2)$, where $p_t(\xi) = n_t/N_t$. Thus in the absence of selection and assuming no differential fertility between mating pairs, the equation of motion is

$$p_{t+1}(\xi) = \iint L(\xi|\xi_1,\xi_2)p_t(\xi_1)p_t(\xi_2)d\xi_1 d\xi_2 \quad (6\text{-}3)$$

or, in terms of the density $n_t(\xi) = N_t p_t(\xi)$,

$$n_{t+1}(\xi) = \frac{r}{N_t}\iint L(\xi)n_t(\xi_1)n_t(\xi_2)d\xi_1 d\xi_2 \quad (6\text{-}4)$$

where $r = N_{t+1}/N_t$ is the net reproductive rate, which we have assumed is independent of ξ. Assortative mating and differential fertility can be included into the model by defining a "mating function" $\varphi(\xi_1,\xi_2)$ (Rocklin and Oster, op. cit.) defined to be the mean number of offspring arising from matings between parents of type ξ_1 and ξ_2. Then,

$$p_t(\xi) = \frac{1}{r_t}\iint L(\xi)\varphi(\xi_1,\xi_2)p_t(\xi_1)p_t(\xi_2)d\xi_1 d\xi_2 \quad (6\text{-}5)$$

where r_t, the mean number of offspring per parent is now given by

$$r_t = \iint \varphi(\xi_1,\xi_2)p_t(\xi_1)p_t(\xi_2)d\xi_1 d\xi_2$$

Finally, selection is introduced by defining a survivorship function, $S_t(\xi)$ giving the fraction of newborn of type ξ which survive to reproduce. The equation of motion now becomes:

$$n_{t+1}(\xi) = \frac{\iint L(\xi)\varphi(\xi_1,\xi_2)S_t(\xi_1)n_t(\xi_1)S_t(\xi_2)n_t(\xi_2)\,d\xi_1 d\xi_2}{\int S_t(\xi)n_t(\xi)d\xi} \quad (6\text{-}6)$$

The definition of the net reproductive rate is now

$$r_t = \frac{\iint \varphi(\xi_1,\xi_2)S_t(\xi_1)n_t(\xi_1)S_t(\xi_2)n_t(\xi_2)d\xi_1 d\xi_2}{\left[\int S_t(\xi)n_t(\xi)d\xi\right]^2} \quad (6\text{-}7)$$

We note that r_t may also be a function of the physiological variable x employed earlier to account for the variations in nutritional state <u>within</u> a particular generation. In general, the interaction between environmental and genetic forces is quite complex; our treatment here must be regarded as only a first approximation.

The equation for the total population is obtained by integrating over all ξ:

$$N_{t+1} = \frac{\iint \varphi(\xi_1,\xi_2)S_t(\xi_1)S_t(\xi_2)n_t(\xi_1)n_t(\xi_2)d\xi_1 d\xi_2}{\int S_t(\xi)n_t(\xi)d\xi} \quad (6\text{-}8)$$

6-3   In order to go further we must introduce specific assumptions about mating and survivorship. The trait we are concerned with here is the genetic component of the dependence of fecundity on protein. Recall that the birthrate function $b(\cdot)$ contained parameters, one of which controlled the threshold for egg-laying. This parameter, heretofore considered fixed, we must now regard as an average with respect to $p_t(\xi)$. Thus it is clear how the changing genetic structure can drive the system through successive bifurcations by altering the effective parameter values (Oster, Iapktchi and Rocklin, 1976).

6-4   For illustrative purposes we shall make the following reasonable assumptions about $S_t(\cdot) \& \varphi(\cdot)$

$$S_t(\xi) = \hat{S}(\xi) e^{-N_t/\hat{K}} \qquad (6\text{-}9a)$$

$$\varphi(\xi_1, \xi_2) = \frac{B}{1 + \gamma |\xi_1 - \xi_2|} \qquad (6\text{-}9b)$$

Substituting into the equation of motion we obtain for the equation for the total population

$$N_{t+1} = N_t e^{r_t (1 - N_t/K_t)} \qquad (6\text{-}10)$$

where the effective net reproductive rate is given by

$$r_t = \ln B + \ln \left[\frac{\iint \frac{d\xi_1 d\xi_2}{1 + \gamma |\xi_1 - \xi_2|} \hat{S}(\xi_1) \hat{S}(\xi_2) p_t(\xi_1) p_t(\xi_2)}{\int \hat{S}(\xi) p_t(\xi) d\xi} \right] \qquad (6\text{-}11)$$

and $K_t = \hat{K} r_t$

Equation (6-10) is a discrete-time logistic model of the type encountered in chapter 4; its behavior under variations in r_t is, by now, familiar. But now, as we see from (6-11) the effective value of r is a functional of the distribution of phenotypes, $p_t(\xi)$. In figure 18 is shown the effect on the bifurcation structure of the variability of the population as measured by the variance, σ_t^2, of $p_t(\xi)$.

One feature of the dynamics of equation (6-10) is worth mentioning. The speed of the system's response to disturbances from equilibrium is faster the greater the genetic variability, as measured by the variance, σ_t^2, of the reproduction kernel. Thus we would expect that a highly inbred strain of flies (i.e. low σ_t^2) would sustain a stable limit cycle longer than the wild type.

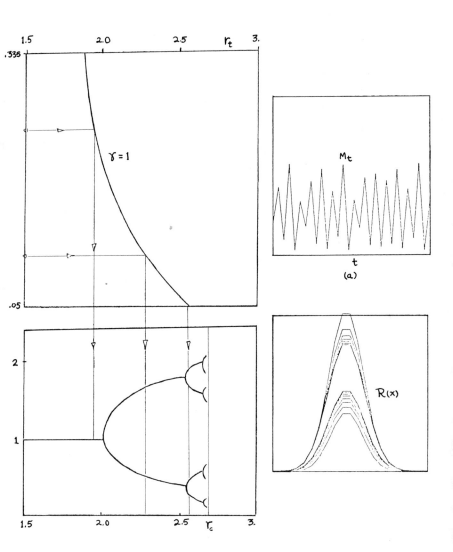

7. Discussion

We have discussed a number of biological phenomena which depend in an essential way on the age and/or phenotype structure of the population. Neither Huffaker's nor

Nicholson's experiments can be understood in terms of models which neglect internal
variables.   In the former case the age-specific nature of the population interaction and
in the latter the competition between age classes were crucial in determining the dynamical behavior of the system.   Moreover, we found that the increased dimensionality of the
models, concommitant with introducing internal variables, introduced new dynamical effects.   Certain features of Nicholson's data could not be understood within the conventional paradigm of equilibria, limit cycles and noise.   The phenomena of cascading bifurcations and strange attractors had to be invoked to explain some aspects of the data.
The notion of "deterministic chaos", characteristic of strange attractors is particularly
intriguing.   In every experiment -- and especially in ecological systems -- there are inevitably stochastic influences which tend to scatter the data about the proposed deterministic mean.   However, we have seen how nonlinearities, acting through the agent of
age and/or phenotype structures can also produce essentially random behavior.   The possibility that strong, but nonlinear, deterministic forces can masquerade as stochastic
"noise" poses some interesting -- and unresolved -- problems for mathematical modelling
in ecology.

BIBLIOGRAPHY

Auslander, D., G. Oster and C. Huffaker. 1974. Dynamics of interacting populations. J. Frank. Inst. 297(5):345-376.

Charlesworth, B. 1973. Selection in populations with overlapping generations. V. Natural selection and life histories. Amer. Nat. 107:303-311.

Crow, J. F. and M. Kimura. 1970. An Introduction to Population Genetics Theory. Harper and Row, New York.

Frederickson, A. G. 1971. A mathematical theory of age structure in sexual populations: random mating and monogamous marriage models. Math. Biosci. 10:117-143.

Guckenheimer, J., G. Oster and I. Ipaktchi. The dynamics of density dependent population models. Theo. Pop. Biol. (to appear).

Hassell, M. and C. B. Huffaker. 1969. Regulatory processes and population cyclicity in laboratory populations of Anagasta kühniella (Zeller) (Lepidoptera: Phycitidae). III. The development of population models. Res. Pop. Ecol. XI:186-210.

Hassell, M. and G. Varley. 1969. New inductive population model for insect parasites and its bearing on biological control. Nature 223:1133-1137.

Hood, L., J. Campbell and S. Elgin. 1975. The organization, expression and evolution of antibodies and other multi-gene families. Annual Review of Genetics, Vol. 9.

Hulburt, H. M. and S. Katz. 1964. Some problems in particle technology. Chem. Ent. Sci. 19:555.

Jacquard, A. 1974. The Genetic Structure of Populations. Springer-Verlag, New York.

Li, T-Y. and J. A. Yorke. 1975. Period three implies chaos. Amer. Math. Monthly (to appear).

Marsden, J. 1973. The Hopf bifurcation for nonlinear semigroups. Bull. Amer. Math. Soc. 79(3):537-541.

Marsden, J. and M. McCracken. 1976. The Hopf Bifurcation. Lecture Notes in Mathematics, Springer-Verlag, New York.

May, R. and G. Oster. 1975. Bifurcations and dynamic complexity in simple ecological models. Amer. Nat. (to appear).

Nicholson, A. J. 1957. The self adjustment of populations to change. Cold Spring Harbor Symp. Quant. Biol. 22:153-173.

Oster, G., D. Auslander and T. Allen. 1975. Deterministic and stochastic effects in population dynamics. J. Dyn. Sys., Meas. and Control (to appear).

Oster, G. and J. Guckenheimer. 1976. Bifurcation behavior of population models. In: The Hopf Bifurcation. J. Marsden and M. McCracken (eds.), Lecture Notes in Mathematics, Springer-Verlag, New York.

Oster, G., A. Ipaktchi and S. Rocklin. 1976. Genetic structure and bifurcation behavior of population models. Theo. Pop. Biol. (to appear).

Oster, G. and Y. Takahashi. 1974. Models for age specific interactions in a periodic environment. Ecol. Monogr. 44:483-501.

Oster, G. 1976. Lectures in population dynamics. In: Modern Modelling of Continuum Phenomena. R. DiPrima (ed.), Amer. Math. Soc.

Podoler, H. 1974. Effects of intraspecific competition in the Indian meal moth on the populations of the moth and its parasite Nemeritis canescens. J. Anim. Ecol. 43: 641-651.

Podoler, H. 1974. Analysis of life tables for a host and parasite (Plodia - Nemeritis) ecosystem. J. Anim. Ecol. $\underline{43}$:653-670.

Rocklin, S. and G. Oster. 1975. Competition between phenotypes. J. Math. Biol. (to appear).

Rotenberg, M. 1972. Theory of population transport. J. Theo. Biol. $\underline{37}$:291-303.

Roughgarden, J. 1972. Evolution of niche width. Amer. Nat. $\underline{106}$:683-718.

Ruelle, D. and F. Takens. 1971. On the nature of turbulence. Comm. Math. Phys. $\underline{20}$: 167-192.

Sinko, J. and W. Streifer. 1967. A new model for age-size structure of a population. Ecology $\underline{48}$:910-918.

Slatkin, M. 1970. Selection and polygenic characters. P.N.A.S. $\underline{66}$:87-93.

Slatkin, M. and R. Lande. 1975. Niche width in a fluctuating environment-density independent model. Amer. Nat. (to appear).

Stein, P. and S. Ulam. 1964. Non-linear transformation studies on electronic computers. Rozprawy Matematyczne $\underline{39}$:401-484.

Von Foerster, H. 1959. Some remarks on changing populations. In: The Kinetics of Cellular Proliferation. F. Stohlman, Jr. (ed.), Grune and Stratton, New York. pp. 382-407.

Weiss, G. H. 1968. Equations for the age structure of growing populations. Bull. Math. Biophys. $\underline{30}$:427-435.

Williams, F. M. 1971. Dynamics of microbial populations. In: Systems Analysis and Simulation in Ecology, I. B. Patten (ed.), Academic Press, New York.

Lectures on Mathematics in the Life Sciences
Volume 8, 1976

NATURAL SELECTION IN AGE-STRUCTURED POPULATIONS

Brian Charlesworth

ABSTRACT

This paper reviews current theory on the process of natural selection in
age-structured populations. The construction of a model of natural selection
in such populations is described and difference equations for a discrete age-
class model are derived. The equilibrium properties of these equations are then
analysed, and a fitness measure for determining equilibrium allele frequencies
is proposed. Various approaches to the dynamical aspects of selection are
outlined. Some unsolved problems are listed.

I. INTRODUCTION

Population genetics is concerned with the dynamic and equilibrium properties
of the genetic composition of populations with Mendelian inheritance, when they
are acted on by evolutionary factors such as mutation, migration, natural
selection, and the stochastic effects of finite population size. The standard
results of theoretical population genetics, recently reviewed by Crow and Kimura
[11], have been derived either on the assumption that generations are discrete,
so that all reproductively mature individuals in any one generation can be
considered to be of the same age, or using continuous-time models which neglect
a_{ς}'-structure i.e. which assume age-independent birth and death rates for each
member of the population. Since populations of many higher organisms, including
man and Drosophila, have overlapping generations with reproduction confined to

---

AMS (MOS) subject classification (1970). Primary 92A10; Secondary 39A10;

45M10.

one period of the life cycle, it is clearly desirable to investigate the extent
to which the conclusions of standard theory can be applied when age-structure is
taken into account. Curiously enough, although this question was given attention
in the early days of population genetics, by Norton [24] and Haldane [14], it has
since been neglected until quite recently, when a number of studies have been
published. These include my investigation of the approach to genetic equilibrium
when evolutionary factors are not operating [6] and Felsenstein's [12] and Hill's
[17] studies of the effects of finite population size in age-structured populations.
Deterministic computer studies of natural selection with age-structure have been
published by Anderson and King [1, 22] and by Giesel and myself [9, 10].
Analytic deterministic results on selection have been obtained by Haldane [15],
Pollak and Kempthorne [25, 26] and by myself [2, 4, 5, 7, 8]. In this paper I
shall review work on the last mentioned topic.

II. CONSTRUCTION OF A DEMOGRAPHIC MODEL OF SELECTION

This topic has been considered in some detail by Norton [24], Pollak and
Kempthorne [26] and by myself [2, 4, 8]. Here I shall present an outline of my
own approach, clarifying some ambiguities in my earlier papers. I shall consider
only a diploid, sexually reproducing population, segregating for n alleles A_1,
A_2 .... A_n at a single gene locus. We thus have to consider $n(n+1)/2$ genotypes
$A_i A_j$ (i,j=1,2..n), each with its own life history schedule characterised by age-
specific survival and reproduction rates. I shall assume initially a discrete-
time demographic model with age-classes 1, 2 .., x, .., d, where d is the index
of the last fertile age-class of that genotype which continues to reproduce
longest. (Only individuals of pre-reproductive or reproductive age need be
considered from the standpoint of population genetics.) In general terms, the
population is characterised by the distribution of numbers of individuals with
respect to age, sex and genotype, present at discrete time-intervals 0, 1, .., t,
.. . Offspring are produced by the set of individuals alive at time t according
to the prevailing age-specific fecundities and the rules of genetics, and pass
into age-class 1 at time t+1. Individuals present in age-class x at time t
either die or pass into age-class x+1 at time t+1, with probabilities which are

in general dependent on their sex, age and genotype, as well as on t.

The following functions are needed. Let l_{ij} (t,x) be the probability that an $A_i A_j$ female conceived at time t-x ($t \geqslant x$) will survive to time t when she will be aged x. Let l_{ij}^* (t,x) be the corresponding male function. Let m_{ij} (t,x) be the expected number of daughters produced by a surviving $A_i A_j$ female aged x at time t, and let m_{ij}^* (t,x) be the corresponding function for sons produced by males. (The use of these fertility functions requires assumptions about sex-ratio, random mating, etc.- see below). Write

(1) k_{ij} (t,x) = l_{ij} (t,x) m_{ij} (t,x),

(2) k_{ij}^* (t,x) = l_{ij}^* (t,x) m_{ij}^* (t,x),

The most convenient way of studying the process of change of total population size is by means of the sequence of values of the variable B(t), which represents the total number of newly conceived individuals produced by the population of individuals present at time t. Similarly, the process of change in the genetic composition of the population can be studied in terms of the sequences of values of the vectors p(t) = $[p_1(t), p_2(t), ..., p_n(t)]$ and $p^*(t) = [p_1^*(t), p_2^*(t), .. p_n^*(t)]$, representing the allele frequencies among the maternally and paternally derived genes, respectively, of the new individuals produced at time t.

Having defined these quantities, it is possible to derive difference equations of order d for the state of the population at a given time t. The following properties of the population are assumed.

1. The sex ratio is constant and independent of parental age and genotype.

2. Mating is at random with respect to both age and genotype.

Random mating with respect to genotype is a standard and usually realistic feature of selection models [11]. Randomness of mating with respect to age is consistent with two different models of the mating process [2]. The first postulates that males shed their gametes into a pool from which they are drawn at random to unite with the eggs produced by females. In such a model we must take into account that males of different genotypes and ages will have different probabilities of contributing gametes to the pool. The model is appropriate for most plant species and for many types of marine animal. The second model assumes that males and females enter a mating pool in each time interval, where pairs are formed by a random collision process. Again, individuals of different genotypes and ages may differ in their probabilities of entering the mating pool. This model is appropriate for many species of animals, but can be considered only as an approximation in the case of species such as man, where partners are selected partly on the basis of the age relations of the two mates. The two models have identical consequences [4], and will be considered together.

These two assumptions enable one to obtain an equation for $p(t)$, assuming $t \geqslant d$ so that all individuals in the population were conceived after time $t=0$. (The modifications necessary for $t < d$ are straightforward [9, 24] and need not be considered here.) For allele $A_i (i=1,2...,n)$ we have

$$(3) \quad B(t)p_i(t) = \sum_{jx} B(t-x)k_{ij}(t,x)[p_i(t-x)p_j^*(t-x)+p_i^*(t-x)p_j(t-x)] ,$$

where summation is understood to extend over all possible values of x and j. Noting that $\sum_i p_i = 1$, we also have

$$(4) \quad B(t) = \sum_{ijx} B(t-x)k_{ij}(t,x)[p_i(t-x)p_j^*(t-x)+p_i^*(t-x)p_j(t-x)]$$

Obviously only n-1 of the set of equations (3) and (4) are independent. Similar equations can be derived for $p^*(t)$ by substituting $k_{ij}^*(t,x)$ for $k_{ij}(t,x)$.

To reduce these equations to a level of simplicity which enables deductions about the dynamics of the system to be made requires further assumptions.
3. The age-specific survival probabilities of males and females of the same genotype are equal, so that

$$l_{ij}^*(t,x) = l_{ij}(t,x).$$

4. The number of offspring produced by a mating between a given pair of individuals is determined solely by the $m_{ij}(t,x)$ function of the female concerned and is independent of the age and genotype of the male.
5. This assumption is the counterpart of assumption 3, but refers to fertility rather than viability, and is thus considerably more complex. Let $M_{ij}(t,x)$ be the expected number of matings in which an A_iA_j male aged x participates during time-interval t. We assume that $M_{ij}(t,x)$ for a given t is a function of i, j and x with the same shape as the female fecundity function $m_{ij}(t,x)$, so that for each i, j, x we can write

$M_{ij}(t,x) = f(t)m_{ij}(t,x),$

where $f(t)$ is some function of t alone, or is a constant. In other words, we are assuming that the probabilities of males of different genotypes and ages mating during any time-interval, relative to the probability for a male of some standard genotype and age, are the same as the fecundities of females relative to a corresponding standard female. The realised fertilities of males are not determined by $M_{ij}(t,x)$ alone, however, but depend on the distribution of females with respect to age and genotype. Using assumption 2, we can write

$m_{ij}^{*}(t,x) = g(t)M_{ij}(t,x),$

where $g(t)$ is some function of t, determined by the state of the female population.

This completes the list of assumptions necessary for simplifying equations (3) and (4). The necessity of imposing severe restrictions on the model of the mating process, in order to get simple equations, was first brought out clearly by Pollak and Kempthorne [26]. They in fact went further than the above list, and assumed that the fecundity of all fertile individuals was the same; this is compatible with the model outlined above, but is not a necessary feature. The use of these assumptions gives the convenient result that, after all individuals present at time t=0 have ceased to breed (t⩾d), the allele frequencies in the gametes produced by the two sexes are equal, so that we have $p(t)=p^{*}(t)$ for t⩾d. This can be seen as follows. Assumption 3 means that the distribution of numbers of individuals with respect to genotype will become the same for males and females born after t=0, so that for times t⩾d the distributions for both genotype and age are the same in the two sexes. It follows from assumptions 4 and 5 that the maternal and paternal allele frequencies among gametes produced by these individuals are the same. When this state has lasted for a number of time intervals greater than or equal to d (t⩾2d), the equivalents of equations (3) and (4) therefore become

(5) $\quad B(t)p_{i}(t) = \sum_{jx} B(t-x)p_{i}(t-x)p_{j}(t-x)k_{ij}(t,x)$

(6) $\quad B(t) = \sum_{ijx} B(t-x)p_{i}(t-x)p_{j}(t-x)k_{ij}(t,x)$

These equations are generalisations of the standard discrete generation equations.

(7) $\bar{w}(t)p_i(t) = p_i(t-1)\sum_j p_j(t-1)w_{ij}(t)$

(8) $\bar{w}(t) = \sum_{ij} p_i(t-1)p_j(t-1)w_{ij}(t)$

where $w_{ij}(t)$ is the fitness of A_iA_j in generation t i.e. the number of progeny that a new A_iA_j individual produced in generation t-1 is expected to contribute to the next generation.

Although several assumptions have had to be made in deriving equations (5) and (6), almost equally severe assumptions lie behind equations (7) and (8) (for discussion of these assumptions see [18], [27], [28]). Equations (5) and (6) thus provide a useful basis for evaluating the consequences of age-structure for selection theory, and most of the remainder of this paper will be devoted to their analysis, with reference to more general cases when appropriate. Before passing on to this analysis, we may note that continuous time systems may be modelled similarly, with t and x becoming continuous variables, $k_{ij}(t,x)$ etc. continuous functions of t and x, and with integration over x replacing summation [14, 24, 2, 5, 7].

III. EQUILIBRIUM CONDITIONS

The conditions for genetic equilibrium in age-structured populations have been treated in several papers [2, 4, 8, 9]. Most of the conclusions concerning equilibria can readily be derived using the more general equations (3) and (4), and remain true to a good approximation even if the assumption of randomness of mating with respect to age is relaxed [8]. The equilibria resulting from the interactions between selection and other factors such as mutation and recombination can also be studied [4, 8]. For brevity, however, the present account shall be based on the simpler equations (5) and (6). I shall use $\hat{p}$ to represent equilibrium values of the vector $p(t)$. For those values of i and j which correspond to alleles which remain in the population at equilibrium, equations (5) and (6) take the equilibrium forms

(9) $B(t) = \sum_j \hat{p}_j \sum_x B(t-x)k_{ij}(t,x)$

(10) $B(t) = \sum_{ij} \hat{p}_i \hat{p}_j \sum_x B(t-x)k_{ij}(t,x)$

Equations (9) and (10) may be compared with the equilibrium forms of equations

(7) and (8)

(11) $\hat{\bar{w}} = \sum_j \hat{p}_j \hat{w}_{ij}$

(12) $\hat{\bar{w}} = \sum_{ij} \hat{p}_i \hat{p}_j \hat{w}_{ij}$

Comparison of these two sets of equations shows that the functions $\sum_x B(t-x) k_{ij}(t,x)/$

$B(t)$ play the same role in determining equilibrium allele frequencies in the

age-structured case as do the quantities $\hat{w}_{ij}/\hat{\bar{w}}$ with non-overlapping generations.

Since in the latter case the equilibrium frequencies depend only on the ratios

of the $\hat{w}_{ij}/\hat{\bar{w}}$ for the different genotypes, we can use the correspondence between

the two cases to write the fitness measure for equilibrium populations with age

structure as

(13) $\hat{w}_{ij} = \sum_x B(t-x) k_{ij}(t,x)/B(t)$

This definition implies [4] that the fitness are normalised such that

$\hat{\bar{w}} = \sum_{ij} \hat{p}_i \hat{p}_j \hat{w}_{ij} = 1$; $\hat{\bar{w}}$ in this case is in fact Fisher's [13] reproductive value

for age-class 1 for the equilibrium population in the case when the population size

is growing at a constant geometric rate.

The $\hat{w}_{ij}$ defined in this way can be used to determine the genetic make-up of

equilibrium populations, using the standard equilibrium equations of population

genetics. For example, with only two alleles A_1 and A_2 and provided that $\hat{w}_{11} \neq \hat{w}_{12}$

$\neq \hat{w}_{22}$, we have

(14) $\hat{p}_1 = (\hat{w}_{12} - \hat{w}_{11})/(2\hat{w}_{12} - \hat{w}_{11} - \hat{w}_{22})$

For $\hat{p}_1$ to represent a non-trivial equilibrium gene frequency ($0 < \hat{p}_1 < 1$), we need

either $\hat{w}_{11}$, $\hat{w}_{22} < \hat{w}_{12}$ or $\hat{w}_{11}$, $\hat{w}_{22} > \hat{w}_{12}$. The correspondence between the fitness

measures for equilibrium populations with age structure and the standard fitness

measures of population genetics can be extended to the more complex situations

mentioned at the beginning of this section [2, 4, 8].

Although there exists this correspondence between the discrete and the age-

structured cases, the latter are distinguished by the biologically important

property that the equilibrium fitnesses depend on the weight given by $B(t-x)/B(t)$

to reproduction at age x. This implies that environmental changes which alter the

demography of the population can induce changes in the equilibrium allele

frequencies even if the relative values of the $k_{ij}(t,x)$ functions for different
genotypes remain unchanged, i.e. if the direction and intensity of selection
remain constant at the fundamental level of age-specific survival and reproduction
rates. A shift in p as a result of this sort of demographic change will of
course imply that $p(t)-p(t-1)$ is no longer zero for a population formerly at
equilibrium, so that we may find that demographic changes which are basically
non-specific with respect to genotype produce genetic changes in a population.
In the case of a population moving along a trajectory with time-independent
$k_{ij}(t,x)$ it is even possible to show formally [4] that a constant age-structure
is necessary for $p(t)=p(t-1)$, unless selection differentials are entirely
independent of age so that one can write

(15) $k_{ij}(x) = c_{ij}k(x)$

where $k(x)$ is common to each genotype, and c_{ij} is a constant characteristic of
A_iA_j. In view of this dependence of genetic equilibrium on demographic stability,
one can simplify equation (13) by writing $B(t-x)/B(t)=e^{-\hat{r}x}$, where $\hat{r}$ is the
geometric rate of growth of the equilibrium population. We then have

(16) $\hat{w}_{ij} = \sum_x e^{-\hat{r}x}k_{ij}(x)$

where $k_{ij}(x)$ is written as independent of t to emphasise that dependence on t
will generally imply changes in the demographic structure of the population.

    Giesel and I [9] have analysed the influence of demographic factors on the
$\hat{w}_{ij}$ as given by the above equations. We found that the age-structure of the
population is the important variable. Demographic changes which alter population
growth-rate but not age-structure, such as age-independent mortality changes, do
not affect the relative fitnesses of different genotypes, whereas changes in
age-structure can change fitnesses even if the population growth-rate remains
constant. In addition, selective systems in which genotypes differ in the
timing of the initiation of reproduction are most sensitive to demographic
changes, whereas systems with differences in age-specific survival probabilities
are very insensitive.

    An obvious problem in using equations like (16) is that $\hat{r}$ is itself a
quantity which is determined by the equilibrium conditions according to the

equation $\hat{w}=1$. When empirical estimations of genotypic fitness are being made, $\hat{r}$ can be calculated from population data [8]. A complete theoretical treatment can be given in the case of a locus with two alleles, A_1 and A_2, and with time-independent $k_{ij}(x)$ functions. In this case, it can be shown [4] that $\hat{r}$ for a population with $0 < \hat{p}_1 < 1$ is the real root of the equation

$$(17) \quad [1 - \sum_x e^{-rx} k_{11}(x)][1 - \sum_x e^{-rx} k_{22}(x)] = [1 - \sum_x e^{-rx} k_{12}(x)]^2$$

Let us define the quantity r_{ij} as the real root of the equation

$$(18) \quad \sum_x e^{-rx} k_{ij}(x) = 1$$

r_{ij} is called the intrinsic rate of increase of $A_i A_j$ and is the asymptotic geometric rate of increase of a population whose members all have the demographic parameters of $A_i A_j$, provided that certain restrictions [19] are placed on the form of $m_{ij}(x)$ as a function of x. It can be shown [3, 4] that $\hat{w}_{11}$, $\hat{w}_{22} < \hat{w}_{12}$ iff r_{11}, $r_{22} < r_{12}$ and $\hat{w}_{11}$, $\hat{w}_{22} > \hat{w}_{12}$ iff r_{11}, $r_{22} > r_{12}$, and that either of these conditions is necessary and sufficient for a unique solution to equation (17), with $\hat{r}$ lying between r_{12} and the nearer of r_{11} and r_{22}. In this case, therefore, the equilibrium properties of age-structured and discrete-generation populations are analogous, since with constant genotypic fitnesses the conditions w_{11}, $w_{22} < w_{12}$ or w_{11}, $w_{22} > w_{12}$ are necessary and sufficient for a unique gene-frequency equilibrium defined by equation (14). Furthermore, we can define a quantity r_p as the real root of

$$(19) \quad \sum_{ij} x e^{-rx} p_i p_j k_{ij}(x) = 1$$

From what has been said above, it follows that if the allele frequency vector is held constant at the value p, the population would stabilize at the constant growth rate r_p. In practice, of course, p will not be constant except at equilibrium, but nevertheless the function r_p proves to be useful. It can be shown [4] that $\hat{r}$ is the maximum value of r_p for r_{11}, $r_{22} < r_{12}$, and the minimum value with the reverse inequality. This can be used to establish a maximisation principle for the process of gene frequency change [7].

An analogous theory can be constructed for the probably more realistic case of a population whose size is regulated by a density-dependent negative feedback mechanism [4]. It is assumed that age-specific survival and/or fecundity rates

at time t for some set R of ages are functions of the vector whose elements are
the total numbers of individuals in some set of age-classes S over a set of earlier
times, T. Write $N(t)$ for the total number of individuals in S at time t, and N_T
for the set of values $N(t)$, $t \epsilon T$. For $t \geqslant d$ we thus have

(20) $N(t) = \sum_{ij, x \epsilon S} B(t-x) p_i(t-x) p_j(t-x) l_{ij}(N_T, x)$

The equations for $p(t)$ and $B(t)$ are now given by equations (5) and (6) with
$k_{ij}(N_T, x)$ in place of $k_{ij}(t, x)$. For a population with $N(t)$ held constant at
some value N we can write k_{ij} as a function of N and x, $k_{ij}(N, x)$. It will be
assumed that for each genotype $\sum_x k_{ij}(N, x)$ is a strictly decreasing function of N.
By analogy with equation (18), this assumption enables us to associate a
parameter N_{ij} with each genotype, which measures the equilibrium population size
of a population all of whose members have the demographic characteristics of
A_iA_j. N_{ij} may be called the carrying-capacity of A_iA_j, and is given as the
(unique) root of

(21) $\sum_x k_{ij}(N, x) = 1$

By analogy with equation (19) we can define N_p as the solution of

(22) $\sum_{ijx} p_i p_j k_{ij}(N, x) = 1$

This represents the equilibrium size of a population with gene frequency held
constant at p, assuming that all populations in the range of values taken by p
would come to a stable size when p is held constant. The analogy with the theory
for time-independent $k_{ij}(x)$ should by now be clear. All the conclusions
associated with equations (17), (18) and (19) can be drawn, substituting N in
place of r. Populations with discrete generations have similar properties [3].

IV. DYNAMICS OF SELECTION

A complete solution of equations (5) and (6) is probably impossible on
account of their non-linearity. Analysis of their non-equilibrium properties has
been concentrated on deriving asymptotic formulae for the ultimate fate of the
population [7, 24, 26], on local analyses of the stability of equilibria [5, and
unpublished], and on difference or differential approximation for the case of
weak selection [7, 14, 15]. I shall consider these three approaches in turn.
Time-independence of the $k_{ij}(t, x)$ functions will be assumed except where otherwise

explicitly mentioned.

1. Asymptotic results. Norton's elaborate analysis of the asymptotic states

reached by two-allele populations under different selection regimes [24] is

probably the most notable contribution to the whole subject of selection with

age-structure. He used integral equation theory and the continuous time

analogues of equations (5) and (6). I have summarised his conclusions in the

right half of Table 1. It will be seen that the intrinsic rates of increase,

r_{ij}, (defined by the integral equation analogues of equation (18)) determine the

ultimate state of the population in a similar way to the w_{ij} fitness measures

with discrete generations. The only major difference lies in the possible

existence of a stable limit cycle about $\hat{p}$ when r_{11}, $r_{22} < r_{12}$.

TABLE 1

The outcomes of various types of selective regimes in two-allele populations

Discrete Generations		Age-Structured Populations	
Selection Regime	Outcome of Selection	Selection Regime	Outcome of Selection
1. $w_{11} = w_{12} = w_{22}$	p_1 constant (neutral equilib.)	$r_{11} = r_{12} = r_{22}$	p_1 constant or neutral limit cycle
2. $w_{11} < w_{12} \leqslant w_{22}$	$p_1 \to 0$	$r_{11} < r_{12} \leqslant r_{22}$	$p_1 \to 0$
3. $w_{11} \leqslant w_{12} < w_{22}$	$p_1 \to 0$	$r_{11} \leqslant r_{12} < r_{22}$	$p_1 \to 0$
4. $w_{11}, w_{22} < w_{12}$	$p_1 \to \hat{p}_1$	$r_{11}, r_{22} < r_{12}$	$p_1 \to \hat{p}_1$, or stable limit cycle about $\hat{p}$
5. $w_{11}, w_{22} > w_{12}$	$p_1 \to 1$ or $p_1 \to 0$	$r_{11}, r_{22} > r_{12}$	$p_1 \to 1$ or $p_1 \to 0$

Pollak and Kempthorne [26] have made a similar analysis of some special

cases of the discrete age-class model for the more restricted situation when

the direction of selection is independent of age, so that when we compare two

genotypes $A_i A_j$ and $A_k A_l$ with $r_{ij} > r_{kl}$, $k_{ij}(x) \geqslant k_{kl}(x)$ with strict inequality for

at least one x. (I have also examined this type of selection using the

continuous model [7]). The results agree with Norton's. Pollak and Kempthorne

also showed that in the case $k_{11}(x) = k_{22}(x)$ and r_{11}, $r_{22} < r_{12}$, p_1 converges to $\frac{1}{2}$.

2. Local analyses (two alleles). In the case of an A_2 allele introduced into a

$\hat{p}_1 = 1$ population, it is very simple [5] to linearise equations (4) and (5) to obtain for the continuous model and with large enough t

(23) $p_2(t) = \int_0^d p_2(t-x) e^{-r_{11}x} k_{12}(x) dx + 0(p_2^2)$

Using standard results of renewal theory [16] , it follows that $\dot{p}_2$ is asymptotical given by

(24) $\dot{p}_2 \sim (r_{12}-r_{11})p_2 + 0(p_2^2)$

If A_2 is recessive to A_1 ($r_{12}=r_{11}$), equation (24) is uninformative. Providing that this is not the case, a mutant gene will spread if it has an increased intrinsic rate when heterozygous, compared with that of the population into which it is introduced, and its rate of increase while still rare becomes asymptotical proportional to its effect on the intrinsic rate of increase. The same analysis can be applied to the equilibrium $\hat{p}_2 = 1$, substituting p_1 for p_2 and r_{22} for r_{11} in equation (24). Putting the two together, it is apparent that a population will tend to maintain two alleles at intermediate frequencies when r_{11}, $r_{22} < r_{12}$, and eliminate both when r_{11}, $r_{22} > r_{12}$, in agreement with Table 1.

A somewhat similar analysis can be carried out for the density-dependent model described in the previous section [5]. It is found that an allele A_i can enter an $A_j A_j$ population if $N_{ij} > N_{jj}$, so that an intermediate gene frequency is maintained if N_{11}, $N_{22} < N_{12}$.

The local stability analysis of equilibria with $0 < \hat{p}_1 < 1$ is more complex and, as far as I know, a complete solution has not been obtained. Linearisation of the two-allele integral equation versions of equations (5) and (6) can be carried out [unpublished results], around $p_1 = \hat{p}_1$ and $B/B = \hat{r}$ as given by equations (14) and (17). It is convenient to define the quantities

$y_1(t) = p(t) - \hat{p}$,

$y_2(t) = \dot{B}(t)/B(t) - \hat{r}$,

and their Laplace transforms with kernel s

$\bar{y}_i = \int_0^\infty e^{-st} y_i(t) dt$ (i=1,2)

After considerable algebraic reduction, we obtain the linear equations

(25) $\begin{aligned} (1-\bar{K}_1)\bar{y}_1 &= f_1(s) - s^{-1}\hat{p}_1(1-\bar{K}_1)\bar{y}_2 \\ (\bar{K}_2-1)\bar{y}_1 &= f_2(s) - s^{-1}\hat{p}_2(1-\bar{K}_2)\bar{y}_2 \end{aligned}$

where $f_1(s)$ and $f_2(s)$ are two integral functions of s determined by the initial
conditions, and where

$$\bar{k}_i = \int_0^\infty e^{-(s+\hat{r})x} [\Sigma \hat{p}_j k_{ij}(x)] dx$$

$$\bar{K}_i = \int_0^\infty e^{-(s+\hat{r})x} [k_i(x) + \hat{p}_i \{k_{ii}(x) - k_{ij}(x)\}] dx \qquad (i,j=1,2)$$

Equations (25) give the following equation, whose roots are the poles of $\bar{y}_1$
and $\bar{y}_2$

(26) $\hat{p}_1(1-\bar{k}_{11})(1-\bar{k}_2) + \hat{p}_2(1-\bar{k}_{22})(1-\bar{k}_1) = 0$

where $\bar{k}_{ii} = \int_0^\infty e^{-(s+\hat{r})x} k_{ii}(x) dx$.

Note that common roots of $1-\bar{k}_1$ and $1-\bar{k}_2$, such as s=0, are excluded from the
poles of $\bar{y}_1$, and common roots of $1-\bar{k}_1$ and $1-\bar{k}_2$ are excluded from the poles of
$\bar{y}_2$.

Using the methods outlined by Lopez [23], it follows from equations (19)
and (20) that for large t we have

(27) $y_1(t) \propto c e^{s_0 t}$

where s_0 is the root of equation (26) with the largest real part (it is assumed
that this is a simple root; suitable modifications can be made if it is multiple).

It is possible to show that when r_{11}, $r_{22} < r_{12}$, equation (26) has no positive
real root. It has not proved possible, however, to exclude the existence of
complex roots with positive real parts in this situation, except in the special
cases $k_{11}(x) = k_{22}(x)$ ($\hat{p}_1 = \frac{1}{2}$) or with selection independent of age, as in equation
(15). The fact that the equilibria $\hat{p}_1 = 0$ and $\hat{p}_1 = 1$ are both unstable implies that,
if complex roots with positive real parts could exist, a stable limit cycle about
$\hat{p}$ would be set up. The existence of such a limit cycle would be of great
biological interest, but I have so far been unable to find a numerical example
which generates one. The analysis of this equilibrium therefore yields a
similar result to Norton's (Table 1), although the methods are very different.

3. Approximations for weak selection. It can be shown [7] that, when selective
differences between genotypes are small, a useful approximate equation for the
vector $\dot{p}$ can be derived in the continuous time, n allele case. An outline of
the argument will be given here. When selection is weak, the absolute value of

the difference in some demographic parameter between a pair of genotypes
(relative to its value for a standard, reference genotype) is less than some
small quantity ϵ. Furthermore $|\dot{p}|$ will be of order ϵ. An adequate
approximation to $\dot{p}$ will thus be obtained by neglecting terms $o(\epsilon)$. The first
step is to find an approximate expression for $B(t)$. Consider a time t_o previous
to the interval $(t-d, t)$. Let the value of $p(t)$ at $t=t_o$ be p_o, and let $p(t-x) = p_o+\Delta p(t-x)$ $(x\leqslant d)$, where $|\Delta p(t-x)|$ is $O(\epsilon)$. From the integral equation version of
equation (6) we have for $t\geqslant d$

$$B(t) = \int_0^d B(t-x) \left[\sum_j p_{io} p_{jo} k_{ij}(x)\right]$$
$$+ 2\int_0^d B(t-x) \left[\sum_i \Delta p_i(t-x) \sum_j p_{jo} k_{ij}(x)\right]dx + o(\epsilon)$$

But we have

$$\sum_j p_{jo} k_{ij}(x) = \sum_{ij} p_{io} p_{jo} k_{ij}(x) + O(\epsilon),$$
$$\sum_i \Delta p_i(t-x) = 0$$

so that

$$(28) \quad B(t) = \int_0^d B(t-x) \left[\sum_{ij} p_{io} p_{jo} k_{ij}(x)\right]dx + o(\epsilon)$$

Using renewal theory, we thus have for large enough $t-t_o$

$$(29) \quad B(t-x) = B(t)e^{-r_o x} + o(\epsilon)$$

where r_o is the real root of the integral equation analogue of equation (18)
with $p=p_o$.

Approximating $p(t-x)$ by $p(t)-x\dot{p}(t)$ and carrying out some algebraic
reduction, we finally obtain by substituting from equation (29) into equation
(5) the following equation, which is accurate to $o(\epsilon)$,

$$(30) \quad \dot{p}_i = p_i \sum_j p_j r_{ij} - \sum_{jk} p_j p_k r_{jk}$$

It can further be shown that $r(t)=\dot{B}(t)/B(t)$ is given to $o(\epsilon)$ by

$$(31) \quad r = \sum_{ij} p_i p_j r_{ij}$$

and that $\dot{r}$ is given to $o(\epsilon^2)$ by

$$(32) \quad \dot{r} = 2\sum_i p_i \left(\sum_j p_j r_{ij} - \sum_{kl} p_k p_l r_{kl}\right)^2$$

Equations (30) - (32) are of the same form as the continuous-time equations of
Fisher [13] and Kimura [20, 21] which do not take age-structure into account.

The properties of these equations are reviewed by Crow and Kimura [11]. Equation (32) is the equivalent of Fisher's Fundamental Theorem of natural selection, and implies that the rate of population growth is maximised by selection, under the assumptions of the model. The equilibrium and stability properties of equations (30) are similar to those of the discrete generation equations (7), substituting r_{ij} for w_{ij}. Furthermore, we can assimilate the equilibrium frequencies predicted by equations (30) to those predicted using the fitnesses given by equation (16) by noting that, with weak selection, we have $r_{ij} \approx C(w_{ij}-1)$, where C is a factor common to all genotypes [7]. Finally, we may note that similar approximate equations can be derived for the discrete age-class model, substituting $p(t)-p(t-1)$ for $\dot{p}$, etc.

A similar theory can be worked out for the model with regulation of population size described earlier. In place of equation (28) we now have

$$(33) \quad B(t) = \int_0^d B(t-x)\left[\sum_{ij}p_{io}p_{jo}k_{ij}(N_T,x)\right]dx + o(\varepsilon)$$

where $k_{ij}(N_T,x)$ is now a continuous functional of the values of $N(t)$ in the interval $(t-d,t)$. Given the assumption made earlier, that a population with constant allele frequencies comes to an equilibrium population size, this equation implies that, to $o(\varepsilon)$ and for t sufficiently greater than t_o, $\dot{B}(t) \rightarrow 0$ and $N(t) \rightarrow N_{po}$. One can approximate N_{po} by $N_{p(t)}$ to $o(\varepsilon)$, so that the allele frequency equation becomes approximately

$$(34) \quad B(t)p_i(t) = \int_0^d B(t-x)p_i(t-x)\left[\sum_j p_j(t-x)k_{ij}(N_{p(t)},x)\right]dx$$

In the two allele case, analysis of this equation shows [unpublished] that the N_{ij} control the final fate of the population in the same way as the w_{ij} in Table 1, with a stable equilibrium $0<\hat{p}_1<1$ when N_{11}, $N_{22}<N_{12}$.

Haldane [14, 15] derived approximate equations for the rate of change in frequency of a dominant allele in a two-allele population with a low rate of growth. Study of his results shows that they may be regarded as approximations to the corresponding versions of equation (30).

V. CONCLUSIONS

From the results which I have outlined above, one can draw the following conclusions about the process of selection in age-structured populations.

1. When selection is weak and there is random mating with respect to age and genotype, with time-independent age-specific survival probabilities and fecundities which are identical in males and females, the intrinsic rate of increase of a genotype A_iA_j can be used as a measure of its fitness in the standard continuous-time equations, at any rate in single-locus situations. Their use yields equilibria and stability properties analogous to those obtained with the standard discrete generation models, using r_{ij} in place of w_{ij}. (It should be noted that these properties may not hold in general in systems with more than one locus [7]). A straightforward extension of this result can be made to the case when there are selective differences between males and females [7], but (contrary to my earlier statement in [7]) the use of r_{ij} as a fitness measure is of doubtful validity when there is non-random mating with respect to age. When there is density-regulation, but the above assumptions are otherwise unchanged, it is found that the ultimate state of a two-allele population is determined by the genotypic carrying-capacities N_{ij}.

2. Equilibrium genotype frequencies in more general cases can be calculated using the fitness measure of equation (16), which is analogous to the discrete-generation measure. The fitness measure for an age-structured population depends in general on the age-structure of the population, with the consequence that variations in age-structure will cause changes in allele frequencies. The implications of this for observational studies of allele frequencies in natural populations have been discussed elsewhere [9]. Empirical estimates of the fitnesses of genotypes in human populations suggest that this process may be operating at the present time [8]. For example, the fitness of sufferers from the genetic disease multiple polyposis of the colon appears to have changed from approximately 76% of that of normal people to 86%, between 1940 and 1964 in the U.S. population, as a result of changing age-structure.

3. With general selection intensities, but with the other restrictions mentioned above, the ultimate genetic make-up of a population with two alleles is controlled by the genotypic intrinsic rates of increase similarly to the

way that the genotypic fitnesses control the fate of discrete-generation
populations. When r_{11}, $r_{22} < r_{12}$, the possibility that a stable limit cycle can
occur, rather than convergence to the equilibrium $\hat{p}$, has not been excluded,
except in certain special cases or with weak selection.

4. The deterministic rate of spread of a (non-recessive) rare allele is
controlled by its effect on the intrinsic rate of increase. This has certain
implications for the evolution of life histories which I have discussed
elsewhere [5]. A more complete treatment of the fate of a new mutant allele
requires a stochastic analysis; results using branching-process theory have
been obtained by Charlesworth and Williamson [unpublished].

Although, as we have seen, some progress has recently been made in the
theory of natural selection in age-structured populations, there are several
questions which need more study. These include:

a. When there are more than two alleles, is there a unique solution (for a
given set of fixed $k_{ij}(x)$ functions) to the equilibrium equations (9) and (10),
analogous to the solution for the two allele case given by equations (14) and
(17)?

b. When r_{11}, $r_{22} < r_{12}$ in the two allele case with fixed $k_{ij}(x)$ functions, can
there be a stable limit cycle, or does the population always converge to the
equilibrium values of p and r? The answer to this question would require
either a deeper analysis of equation (25) than I have been able to provide, or
a different approach to the stability problem.

c. To what extent do the conclusions concerning the dynamics of selection have
to be modified to take the effects of non-random mating with respect to age
into account?

REFERENCES

1. W.W. Anderson and C.E. King, Age-specific selection, Proc. Natl. Acad. Sci.
   66 (1970), 780-786.

2. B. Charlesworth, Selection in populations with overlapping generations. I.
   The use of Malthusian parameters in population genetics, Theor. Pop. Biol. 1
   (1970), 352-370.

3. B. Charlesworth, Selection in density-regulated populations, Ecology 52 (1971) 469-474.

4. —————————, Selection in populations with overlapping generations. III. Conditions for genetic equilibrium, Theor. Pop. Biol. 3 (1972), 377-395.

5. —————————, Selection in populations with overlapping generations. V. Natural selection and life histories, Amer. Natur. 107 (1973), 303-311.

6. —————————, The Hardy-Weinberg law with overlapping generations, Adv. Appl. Prob. 6 (1974), 4-6.

7. —————————, Selection in populations with overlapping generations. VI. Rates of change of gene frequency and population growth rate, Theor. Pop. Biol. 6 (1974), 108-133.

8. B. Charlesworth and D. Charlesworth, The measurement of fitness and mutation rate in human populations, Ann. Hum. Genet. 37 (1973), 175-187.

9. B. Charlesworth and J. T. Giesel, Selection in populations with overlapping generations. II. Relations between gene frequency and demographic variables. Amer. Natur. 106 (1972), 388-401.

10. ——————————————————, Selection in populations with overlapping generations. IV. Fluctuations in gene frequency with density-dependent selection. Amer. Natur. 106 (1972), 402-411.

11. J. F. Crow and M. Kimura, An introduction to population genetics theory, Harper and Row, New York, 1970.

12. J. Felsenstein, Inbreeding and variance effective number in populations with overlapping generations, Genetics 68 (1971), 581-597.

13. R. A. Fisher, The genetical theory of natural selection, Clarendon Press, Oxford, 1930.

14. J.B.S. Haldane, A mathematical theory of natural and artifical selection, Part IV, Proc. Camb. Phil. Soc. 23 (1927), 607-615.

15. —————————, Natural selection in a population with annual breeding but overlapping generations, J. Genet. 58 (1963), 122-124.

16. T. E. Harris, The theory of branching processes, Springer-Verlag, Berlin, 1963

17. W. G. Hill, Effective size of populations with overlapping generations, Theor.

Pop. Biol. 3 (1972), 278-289.

18. O. Kempthorne and E. Pollak, Concepts of fitness in Mendelian populations,

    Genetics 64 (1970), 125-145.

19. N. Keyfitz, Introduction to the mathematics of population, Addison-Wesley,

    Reading, Mass., 1968.

20. M. Kimura, Rules for testing stability of a selective polymorphism, Proc.

    Natl. Acad. Sci. 62 (1956), 70-74.

21. ————, On the change of population mean fitness by natural selection,

    Heredity 12 (1958), 145-167.

22. C.E. King and W.W. Anderson, Age-specific selection. II. The interaction

    between r and K during population growth, Amer. Natur. 105 (1971), 137-156.

23. A. Lopez, Problems in stable population theory, Office of Population Research,

    Princeton, N.J., 1961.

24. H.T.J. Norton, Natural selection and Mendelian variation, Proc. Lond. Math.

    Soc. 28 (1928), 1-45.

25. E. Pollak and O. Kempthorne, Malthusian parameters in genetic populations.

    Part I. Haploid and selfing models, Theor. Pop. Biol. 1 (1970), 315-345.

26. ————————————————, Malthusian parameters in genetic populations.

    Part II. Random mating populations in infinite habitats, Theor. Pop. Biol.

    2 (1971), 357-390.

27. T. Prout, The relation between fitness components and population prediction

    in Drosophila. I. The estimation of fitness components, Genetics 68 (1971),

    127-149.

28. ————, The relation between fitness components and population prediction

    in Drosophila. II. Population prediction, Genetics 68 (1971), 151-167.

SCHOOL OF BIOLOGICAL SCIENCES
UNIVERSITY OF SUSSEX
FALMER, BRIGHTON, SUSSEX BN1 9QG
ENGLAND

Lectures on Mathematics in the Life Sciences
Volume 8, 1976

MATHEMATICAL DEVELOPMENTS IN HODGKIN-HUXLEY THEORY AND ITS APPROXIMATIONS

Hirsh Cohen

1. INTRODUCTION

In the past few years, a set of equations which describe the way a voltage pulse travels along a nerve cell has received a great deal of attention from physiological experimenters and some attention from mathematicians. Why do these equations merit interest? What problem has been solved? Is it, in fact, solved and has it created, as interesting problems often do, other questions of physiological or mathematical interest?

One of the two major functions of a single nerve cell is the conduction of information through itself to neighboring cells. (The other function is the processing or treatment of that information but that will not be of primary interest here.) It has been known for many years [1] that voltage pulses deliver the information traveling along the membrane of the axon of the nerve cell, that these pulses travel at near constant velocity, and are of fixed shape including fixed amplitude. If the pulse, or action potential, as it is called, does not arrive at the end of the axon or arrives in distorted condition or if its speed varys greatly, the cell will not be providing one of its major functions, the message will not get through. If a group of cells malfunction in this fashion, a muscle will not respond, or a sensory organ will not get its perception of the outside world successfully into the whole organism. Thus the pulse propagation is an important element of the nerve system.

Mathematically, the equations that are currently used to describe this phenomenon are non-linear diffusion equations. Diffusion equations do not normally support constant velocity, constant shape, wave propagation. The

89

non-linearity must provide the difference in behavior. In the nerve equations,
the non-linearity represents the ionic current flow across the axon membrane;
this is the energy supply that maintains the pulse shape and velocity as it
moves along the membrane and prevents it from decaying, as it would if it were
a linear diffusion phenomenon.

Hodgkin-Huxley theory [2] represents a non-myelinated* axon membrane as a
one-dimensional transmission line. This is a plausible approximation because
1) the membrane is the only part of the axon that has high enough impedance to
support the potential difference change needed for the pulse; 2) the membrane
is very thin (100 angstroms) compared to the radius of the axon (from 5 microns
to 250 microns); 3) the electrical characterization of the membrane can be
achieved fairly accurately in terms of distributed constant capacitance across
it and distributed constant resistance in the inner and outer surrounding
materials.

The remaining required distributed element,

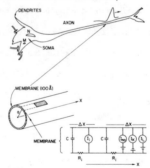

Figure 1.   Schematic representations of a motor nerve cell, its axon, and
                    the distributed circuit for the membrane.

the conductivity across the membrane, is not constant. Indeed, if it were,
constant shape, constant velocity waves of finite amplitude could not be
supported. The conductivity is, in fact, a function of the electric field; it
is related to the transport of, primarily, sodium and potassium ions. In the

*A myelinated axon has a sheath around the membrane that only allows active ion
 transport at spaced nodal sections; a non-myelinated axon allows transport
 anywhere along its length and circumference.

theory of Hodgkin and Huxley, the cross-membrane ionic current is measured experimentally.

If $v(x,t)$ is the voltage across the membrane, C is the fixed capacity, R_i the specific resistance of the material within the membrane, a the axon radius, then the equation for current density is

$$C \frac{\partial v}{\partial t} + I_i(v;Na,K) = \frac{a}{2R_i} \frac{\partial^2 v}{\partial x^2} \tag{1}$$

Note that because the membrane and the axon have no important magnetic properties, there is no inductive term and the transmission line equation has only a first order derivative in time.

Before describing the ionic current term I_i and introducing the new variables required for this description, it may be useful to say what we require of a theory:

i) the equations should exhibit a threshold (figure 2); for a given amplitude of voltage or current stimulation, a minimum duration of this excitation is required before a pulse is formed and is conducted along the axon.

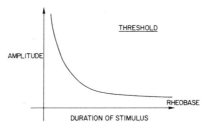

Figure 2.   Schematic of the threshold curve.

ii) depending on initial or boundary data, corresponding to a particular experimental arrangement, the equations should yield stationary or moving pulse solutions. The three conditions are:

a. space clamped - the axon is shorted along its length so that pulses developed across the membrane do not travel. The only variable is time. order equations (3) and therefore the relation to v is fairly complex. The most drastic approximation is to hide all of the kinetics by making I_i a simple function of v only, for example a quadratic or cubic polynomial. The idea of using a cubic comes directly from the behavior of the sodium ion current as

b. propagation - the assumption is made that a pulse has developed and
reached a steady traveling state. The solutions are functions of a traveling
variable, $x + \theta t$, where θ is a constant velocity.

c. initiation - the stimulus is provided at a point on the axon, the
pulse develops as it moves along the axon and eventually reaches the shape and
velocity found in (b). The solutions are functions of x and t independently.

The pulses referred to are typically about 100 millivolts in maximum
amplitude, two milliseconds long, have a steep rising front, a less steep
trailing edge slope. They have the characteristics that the trailing edge
becomes negative for a period long compared to the pulse width.

iii) alterations in geometry, stimulation, chemical environment, or
temperature may change the form of the solution and these changes should
correspond with experiments, e.g.:

a. changes in axon diameter produce changes in pulse velocity and shape

b. temperature changes alter thresholds, pulse velocity and amplitudes

c. changes in sodium or potassium concentrations or substitutions of
other univalent ions can alter pulse velocity and amplitude or remove the pulse
entirely. Changes in calcium concentration also affect the pulse

d. sub-threshold stimuli should produce a non-propagating, decaying
signal; very large super-threshold stimuli may produce more than one pulse (a
finite train) or a very long (infinite) train of pulses.

There are a number of other questions that can be asked, not all of which
have yet been posed experimentally.

In order to produce solutions we need a description of the ionic current.
Hodgkin and Huxley introduced three new variables for this purpose, two of
which m and h are related to the sodium ion transport and the third, n,
to the potassium transport. With these variables

$$I_i = \bar{g}_{Na}\ m^3\ h\ (v-\bar{v}_{Na}) + \bar{g}_K\ n^4\ (v-\bar{v}_K) + \bar{g}_\ell\ (v-\bar{v}_\ell) \tag{2}$$

$\bar{g}_{Na},\ \bar{g}_K,\ \bar{g}_\ell,\ \bar{v}_\ell,\ \bar{v}_{Na}$ and $\bar{v}_K$ are constants. The new variables m, n, and h are
defined by the equations

$$\frac{\partial m}{\partial t} = \frac{1}{\tau_m(v)} \ (m_\infty(v) - m)$$

$$\frac{\partial h}{\partial t} = \frac{1}{\tau_h(v)} \ (h_\infty(v) - h) \tag{3}$$

$$\frac{\partial n}{\partial t} = \frac{1}{\tau_n(v)} \ (n_\infty(v) - n)$$

The six functions τ_m, τ_h, τ_n, m_∞, h_∞, and h_∞ are experimentally derived by a technique called the voltage clamp.* The functions m and n appear to the third and fourth power respectively in equation (2) because this seems to represent a good fit to measured ionic current data. The representation is not unique, however, two new variables instead of three have also been proposed and the powers of m and n could be other than three and four [3]. To have an idea of the form of the six empirically determined functions for the membrane axon of the squid, Hodgkin and Huxley found

$$\frac{1}{\tau_m} = \frac{0.1(25-v)}{[\exp(\frac{25-v}{10})-1]} + 4 \ \exp(\frac{-v}{18})$$

$$m_\infty(v) = \frac{0.1(25-v)\tau_m}{[\exp(\frac{25-v}{10})-1]}$$

with similar expressions for the n and h quantities.

These equations form a fifth order system of partial differential equations. They have been solved for all of the conditions we have mentioned earlier but these solutions have been all numerical. There are, as one might expect no explicit, exact solutions. In what follows, we will exhibit what is known about the numerical results and their utilization in experiments, the qualitative behavior and some approximations.

The approximations have to do with the ionic current. In Hodgkin-Huxley theory it is the linear composition of three currents. The conductivities are dependent on m, n, and h and v but these, in turn, are solutions to the first

---

*In this method, the axon is shorted as in the space clamp by threading an electrode through it but in addition, the voltage across the membrane is held fixed. This allows for measurement of the total ionic current as a function of time. By changing the sodium ion concentrations outside the membrane, sodium and potassium current can be separated.

measured in Hodgkin-Huxley theory. Simple flame propagation, and distributed
chemical reactions of other kinds are governed by equations of this type.

An important observation is that there are fast and slow time scales
associated with the pulse formation.

The function, τ_m, is the "non-linear response time" for the sodium ion
current during the rising phase of the voltage pulse. τ_n and τ_h describe,
respectively, the response times for the potassium ion current and for the
falling phase of the sodium current. τ_m is much smaller than τ_n and τ_h
as can be seen from figure (3). The behavior of m,n, and h as functions of
time in their relationship to v is shown in figure (4) to conform to the
"response times".

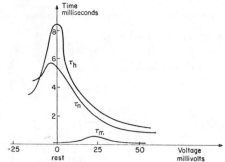

Figure 3.   The non-linear response times of the functions m,n, and h as
functions of the voltage.[*]

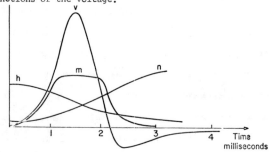

Figure 4.   The time-dependence of m,n, and h related to v as a function
of time.[**]

[*]Reprinted from Electrochemistry in Biology and Medicine (edited by T.
Shedlovsky) p. 121, with permission of John Wiley & Sons, Wiley, N.Y.

[**]Reprinted from Gen. Physiol. 43 (1960), pp. 867-896, with permission of
Rockefeller University Press.

The rising front of the pulse, obtainable from the second order equations with polynomial non-linearity is on the fast scale. The "turning off" of the pulse, its trailing edge and the recovery is on the slow scale. Most of the approximations make use of the difference in these time scales.

In the next step beyond the second order equation one adds the slow kinetics. The simplest form is just a linear equation

$$\frac{\partial v}{\partial t} + f(v;a) + z = \frac{\partial^2 v}{\partial x^2} \qquad (4)$$

$$\frac{\partial z}{\partial t} = \varepsilon v \qquad (5)$$

The third order system allows for not only a solitary pulse but also periodic solutions in the form of infinite trains. For each value of ε and a there is a solution with infinite period. By choosing a finite period (above a certain lower limit), each value of ε and a give a periodic pulse train with lower velocity than for the solitary pulse.

Before embarking in more detail on these main topics it might be useful to mention just what are the physiological problems that remain to be solved. As we will see, the equations given above do represent most of the measured phenomena and can be used to predict new effects for measurement. However, it is a phenomenological theory, essentially a transmission line with an empirical expression for the non-linear shunt conductivity. The key problem now is to develop an understanding of this electric field dependent conductivity, or said another way, the variable membrane permeability and its different sensitivities to the sodium and potassium ions. This requires an understanding of molecular structure, dynamics, and electric charge patterns. This is a hard problem in biophysics but a large amount of experimentation is underway [4] and as the data arrive , theories will become possible.

There are other problems for the single nerve cell that deserve theoretical attention or, at least, soon will. Examples include the signal transmission process from the dendrites of the nerve cell into the beginning of the axon, the chemical transmission of signals across the synaptic gap and the related phenomena of facilitation and depression of transmitted signal amplitudes, and

other processes in the cell such as axonal flows of internal materials which
may or may not be related to the information processes of the cell.

There is, of course, a large field of theoretical inquiry into the
relationship between networks of nerve cells and the signaling and information
carrying conducted by them. One of the basic theoretical problems is to
understand how single nerve cell behavior of the kind discussed here affects
cell population behavior, the sensitivity problem.

2. NUMERICAL RESULTS

While the primary aim of this exposition is to describe the state of the
qualitative understanding of these kinds of equations, it is useful to first
say what has been learned quantitatively through numerical computation. There
has been one set of computations which verify the basic theory: given the ionic
functions, calculate a pulse corresponding to space clamped, steady state propagating, and initial value problem boundary conditions. The second set of computations is more interesting because these create changes in the normal environment and test the theoretical axon's response. The kinds of changes are geometrical, chemical, electrical, and temperature. For some of these calculations there
are experimental counterparts and for others, one hopes experiments will be done.

There is a good deal of discussion in the book of Cole [3] on the question
of calculations that verify experimental observations and also those which
assist in understanding experimental techniques. The numerical methods which
have been used in the calculations described are, for the most part, standard
ones. Discussions are given in [5], [6], [7].

A. Verification (normal conditions)

1. Space clamped

The equations to be solved are

$$C \frac{\partial v}{\partial t} + I_i(v,m,n,h) = I \tag{6}$$

where I is the given stimulus, together with equations (2) and (3) taken now
as ordinary differential equations in t alone. The initial conditions are

$$v(0) = v_0, \qquad m(0) = m_0, \qquad h(0) = h_0, \qquad n(0) = n_0$$
$$v(\pm \infty) = 0, \qquad v(t) \not\equiv 0 \tag{7}$$

Solutions have been carried out using iterative predictor-corrector methods and step-by-step Runge-Kutta [5a]. The results are satisfactorily close to experiments (figure 5)

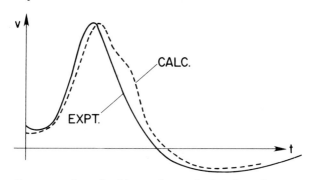

Figure 5.   The space-clamped voltage pulse, calculated results compared to experiment.[*]

2. Propagation

The equations are:

$$v(x,t) = v(x+\theta t), \quad m(x,t) = m(x+\theta t) \text{ etc.} \tag{8}$$
$$x+\theta t = \xi \tag{9}$$

" ' " refers to differentiation with respect to ξ

$$\frac{a}{2R_i} v'' - C\theta v' - I_i(v,m,n,h) = 0 \tag{10}$$
$$\theta m' = M(m,v) \tag{11}$$
$$\theta n' = N(n,v) \tag{12}$$
$$\theta h' = H(h,v) \tag{13}$$
$$v(\pm \infty) = 0, \qquad v \not\equiv 0 \tag{14}$$

where M,N, and H stand for the right hand sides of the equations in (3).

The original calculation of Hodgkin and Huxley was a desk machine "shooting" method, predictor-corrector, which sought a value of θ which provided a non-zero, pulse shaped solution for the fifth order ordinary differ-

*Reprinted with permission from SIAM Journal on Applied Mathematics (1955). Copyright 1955 by Society for Industrial and Applied Mathematics.

ential equation system. This is a "non-linear eigenvalue problem". The pulse
velocity, for squid axon data, was found to be 18.8 m/sec, compared to an
experimental value of 21.2 m/sec, quite satisfactory. A subsequent calculation
by FitzHugh and Antosiewicz by Runge-Kutta [5b] suggested there might be two
pulse solutions available and this was shown to be the case by Huxley [8]
The second pulse has a slow speed and a small amplitude. It is not found
experimentally. The implications of this second solution will be discussed
shortly.

3. Initial value problem

The equations are (1), (2), and (3) with the conditions

$$v(x,0) = 0; \qquad x > 0 \tag{15}$$

$$m(x,0) = 0, \; n(x,0) = n_0, \; h(x,0) = h_0; \; x > 0 \tag{16}$$

$$v(0,t) = V(t), \qquad t > 0 \tag{17}$$

$$v(\pm \infty, t) = 0, \tag{18}$$

By solving this problem one can examine how the local excitation caused
by the stimulus grows, propagates, and finally reaches the steady state solution
described above. This is the most important verification. It was first done
for the complete Hodgkin-Huxley system by Cooley and Dodge [6] (figure 6).

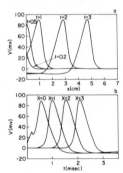

Figure 6.   The generation of a voltage as calculated from an initial-boundary
             value problem for the partial differential equations.[*]

The initial pulse very quickly achieves a constant, conduction velocity, just

*Reprinted from Biophysical Journal (edited by F.A. Dodge, Jr.), Vol. 6,
No. 5 (1966), pp. 583-599, with permission of Rockefeller University Press.

immediately after the local action potential at x = 0 reaches its maximum, the traveling pulse velocity is already within a fraction of a percent of its final value. The initial boundary value problem also verifies the existence and nature of the threshold for the normal axon. For a fixed duration of stimulus, a current amplitude can be found that just sets off a pulse. Figure (7) exhibits the computed data and also shows the space clamped threshold.

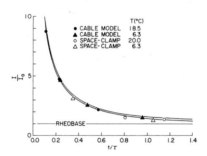

Figure 7.   Verification of the threshold phenomenon by calculation. Both space clamped and propagating cases have been calculated at two different temperatures.[*]

B. Altered environments

1. Space clamped

a. Approach to threshold.

For a fixed, short duration of stimulus, if the amplitude of the injected current is increased, narrowing the gap between a given stimulus and the threshold value, the onset of the pulse and its maximum amplitude change. The pulse takes longer to form and the amplitude decreases (figure 8) [5a].

---

*See footnote to Figure 6, page 98.

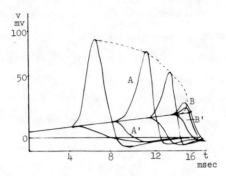

Figure 8.   Response for stimuli just below and just above threshold, showing
               the effect of narrowing the gap between the two. A and A'
               differ by one part in 10^8 and B and B' by one in 10^{14}. *

          b.   Repetitive firing [5b].

     When the current stimulus has a large enough amplitude and is kept on for

a very long time, more than one action potential is generated. This is to be

expected in sensory nerve systems and, in fact, they depend strongly on the

relationship between stimulus amplitude and pulse frequency in such systems,

continuous trains of pulses generally appear with alteration of frequency

indicating change in stimulus. The calculations for the space-clamped squid

axon also produced infinite trains of action potentials for large enough current

stimulus. For these, however, there was very little change in frequency with

change in stimulus. For a small range of stimulus just above the value at

which single pulses are generated, finite trains of two and three pulses can

appear (figure 9). These finite trains are interesting because, since we are

now dealing with an ordinary differential equation system, they represent

nearly repetitive cycles, each lying on or close to the same closed locus, but

gradually moving away until finally one returns to a real equilibrium point.

It should be understood that, although the squid motor axon does produce such

_____

     *See footnote to Figure 5, page 97.

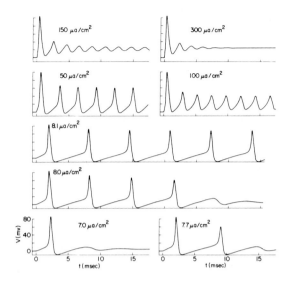

Figure 9.   Repetitive firing for the space clamped axon. The stimulus current
is maintained at the given amplitudes for very long time periods.

short finite trains, it is not a frequently occurring mode in motor neurons.

The fact that Hodgkin-Huxley equations do not provide, in their original
form, repetitive pulses over a large range of stimulus amplitudes nor strong
influence of this amplitude on pulse frequency means that the model is not
really suitable for sensory pulse conduction. Dodge [9] has indicated
that an alteration in the form of the potassium conductance can provide
frequency modulation.

c. Changes in external calcium concentrations

When the concentration of calcium is changed in the bathing solution
external to the axon, there is a shift in the sodium and potassium conductivities. This has the effect of, in turn, shifting the voltage. Small subthreshold oscillations and action potentials can be obtained by controlling

the calcium concentration. Huxley [10] has calculated these effects using
the fact that the shift in $v, \Delta v$, is given by

$$\Delta v = k \ \ell n \ \frac{[C_a]}{[C_a]_{normal}} \tag{19}$$

where $k = 9.32$. The detailed results are given in [10].

Huxley's calculations show that lowering the calcium concentration when
the stimulus would normally only be strong enough to give small oscillations
can change these from being heavily damped to oscillations of exponentially
increasing amplitude. There is also a change in period but not a large one.
Huxley noted that there is a range in which the subthreshold oscillation
amplitude decrement is nearly zero and that just in this region the frequency
of spikes generated as threshold is reached is very close to the subthreshold
oscillation frequency. This observation led to other experiments and some
calculations [11]. For larger stimuli, those which produce action potentials,
changes in calcium concentration can move the response from a single pulse
through multiple pulses to an infinite train of pulses.

d. Effects of periodic stimulation

If a periodic sequence of square current pulses are applied to a clamped
axon a number of phenomena result that are functions of the driving frequency
and amplitude. In the calculations of Berkenblit et al.[12], the duration of
each stimulating pulse was kept fixed at 0.5 msec.

i. Subthreshold

Calculations were made with a subthreshold stimulus ($\sim$40% threshold) and
a period of lmsec. The membrane can be seen to respond passively but the
period is such that the response gradually increases until it reaches threshold
and action potential formation. figure (10) The threshold value does not
depend on the frequency.

ii. Blocking of pulses

It has been known experimentally that if a sequence of superthreshold
stimuli are given at a moderate rate (a period, T, of 10 milliseconds) spike

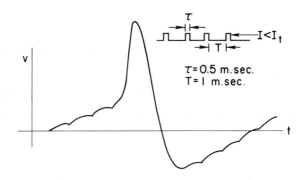

Figure 10.   Response of the clamped axon to subthreshold repeated stimuli.

generation is affected in several ways. Most conspicuous is that, depending

on the frequency and duration of the stimulus, pulses will drop out of the

sequences, as shown in figure (11).

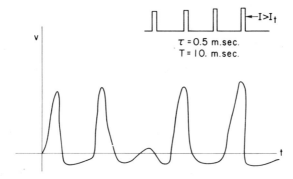

Figure 11.   Response of the clamped axon to superthreshold repeated stimuli.

Berkenblit et al. have calculated blocked pulse trains of this kind in which

the third, fourth, fifth, sixth, or seventh pulse falls out. They use short

stimuli (0.5 msec). They also show how the sodium and potassium conductance

functions respond. Their calculations reveal, as one might expect, that the

m function follows the voltage pulse closely and, in particular, has returned

to near its initial value at the end of each full pulse. The h and n

functions, because of the large values of τ_h and τ_n, respond slowly and
when the stimulus re-occurs, have not returned to their original state. The
function n gradually rises, the function h gradually decreases, until for
some stimulus, they have brought the membrane below threshold. This is, then,
another type of stimulus duration, amplitude and frequency interaction. In
the paper cited [12], a plot is exhibited showing various ranges of blocked
pulses for the associated stimulation amplitude and duration.

iii. High frequency stimulation

In the same report, calculations are made for stimulation at a higher
frequency than would be normal for the membrane. The result is that pulses are
missed, again. For 1000 cycles/second, the actual frequency of response is
lower than for 500 cycles/second: only one pulse in eight or one in seven is
set off at 1000 cycles/second. The "miss rate" varies between these two. For
500 cycles/second it is more like 4:1 and 3:1.

2. Propagation

For the steady propagating pulse, as we have mentioned, at least two
pulses have been calculated as solutions to the fifth order differential
equation system for given chemical, dimensional, and temperature conditions.
If the temperature is altered, a new set of dual solutions can be formed and
Huxley [ ] was able to calculate a region where two pulses might exist. The
calculations show that for high temperature no pulse at all can exist. (Temperature enters the theory through the ion kinetics. If $\phi = 3^{(T-6.3).10}$, where T
is the temperature in degrees centigrade, then the direct effect is to multiply
the right hand side of the equation in (3) by ϕ. This is equivalent, however,
to changing from t to a new time variable and a new capacitance, $t'=\phi t$ and
$C'=\phi C$). Experimental evidence of this calculated effect was given by Hodgkin
and Katz [13].

Similarly Cole and FitzHugh (see [3]) have found "non-unique" solutions
when the leakage conductance, $\bar{g}_L$, is varied. Dodge and Cooley [6] used changes
in the maximum sodium and potassium conductivities, $\bar{g}_K$ and $\bar{g}_{Na}$. Lowered values

were throught to roughly correspond to the effects of narcotics. If both conductivities are attenuated by a factor η, then two values of velocity and
maximum amplitude are found for each value of η, figure (12)

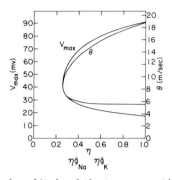

Figure 12.   The speed and amplitude of the two propagating pulse solutions

available as functions of η (reductions of $\bar{g}_{Na}$, $\bar{g}_K$).

Again there is a cutoff for values of $\bar{g}_{Na}$ and $\bar{g}_K$ too small.

Huxley conjectured [8] that the low amplitude, low velocity pulse is the
assympototic state obtained from an initial stimulus which is just subthreshold

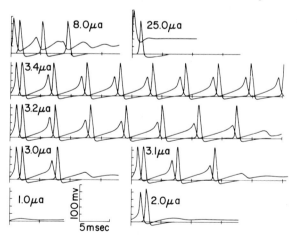

Figure 13.   Repetitive propagated pulses. The stimulus current is maintained

at the given amplitudes for very long time periods.[**]

---

*See footnote to Figure 6, page 98.

**See footnote to Figure 6, page 98.

and that these pulses are unstable, they do not ever attain this asympototic state. We will discuss this aspect further later on.

>   3. Initial boundary value problem for the partial differential
>   equations

a. Repetitive firing

Like the space clamped case, the initial value problem also has solutions with more than one impulse. These are shown in figure (13). Again, as with the clamped case, the range of current where multiple pulses occur is narrower than the real physiological range and, in addition, the desired frequency modulation is missing.

Dodge and Cooley [6] attempted to find the low velocity pulse mentioned in the previous section by solving the initial value problem for the partial differential equations with a subthreshold stimulus but one very close to threshold. The results are shown in figures (14), (15), (16) for several above and below threshold values. For these normal axon calculations, however, the subthreshold pulse does not persist.

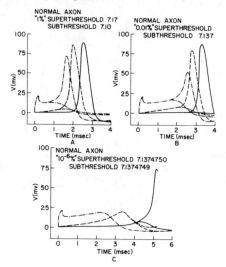

Figure 14. Calculations for the partial differential equation initial-boundary
15.
16.   value problem for super and subthreshold stimulus values.*

---

*See footnote to Figure 6, page 98.

Application of the previously mentioned "narcotizing" factor, η, does have some effect. For $\eta = \frac{1}{3}$, the two pulses all seem to travel together but at reduced amplitude and velocity. The axon is, perhaps, "desensitized and hence the unstable pulse is less unstable. (see [6] for details)

Slow moving pulses which do not have large amplitudes may play a role in propagation along some dendrites from the synapse toward the cell body.

b. Other chemical changes

Another chemical change involves lowering only $\bar{g}_{Na}$. It is known that this is what happens when tetrodotoxin is added to the bathing solution [4]. Khodorov et al. [14] calculated the effect on the propagation pulse of doping a segment of the axon some distance away from the point of stimulation. When this began at 2.025 cm from x=0, with $\bar{g}_{Na}$ reduced by a factor of ten, the effect was felt upstream at x=1.8 cm and the pulse was quite thoroughly diminished in amplitude at 3 cm (figure (17)).

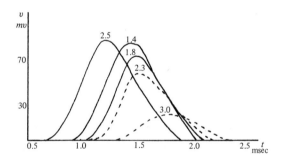

Figure 17. The effect on the pulse of a reduction of $\bar{g}_{Na}$ by a factor of 10. The change begins at x = 2.025 cm. Pulse shapes are shown for the values of x in cm given with each curve.

In another numerical experiment similar to that of Cooley and Dodge described above, Khodorov et al. studied the effect of simultaneously reducing $\bar{g}_{Na}$ and $\bar{g}_K$ at a section of the axon. For a decrease by a factor of 3, the calculations show a decrease in speed as the pulse crosses into the altered zone. Then

there is a small increase in speed in this zone but not enough to return to the
original velocity. For this value of conductivity change there was no upstream
effect on the on-coming pulse (figure (18)). The change again takes place at
2.025 cm. Khodorov et al. have also studied the effects of calcium concentration change on propagation.

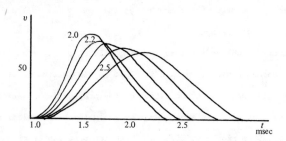

Figure 18. The effect of reducing both $\bar{g}_{Na}$ and $\bar{g}_{K}$ beginning at x = 2.025.
            Numbers on the curves indicate values of x.

c. Geometrical effects

The only geometrical quantity that enters into the equations is the radius.
If there is a change in the radius along the axon, there will be a pulse
velocity change. Khodorov et al. [15] have calculated the effect of step
increases in the radius. The impedance met by the advancing ion current is
altered as the pulse approaches and passes through the step.

When the radius is increased by a factor of five the following changes
occur (the increase is at 2.025 cm from x=0): as the pulse approaches the change
(at 1.8 cm) its amplitude and its initial rate of rise decrease. The rate of
total ion conduction also decreases. Just before the change (at 2.0 cm), the
pulse shape is strongly altered so that it has an initial rise to about a quarter
of the maximum amplitude, a plateau, and then a further rapid rise. There is a
reverse wave of depolarization, a passive reflection, backwards from the expanded
region. This is seen in the pulse shape at 1.8 cm and constitutes a kind of
"echo" effect (figure (19)).

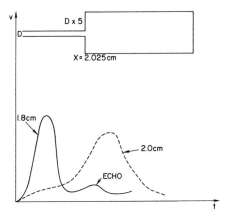

Figure 19. The effect of a change in axon radius on pulse propagation.

The velocity drops in front of the change, jumps up at the point and then rises to a new, higher value.

If the step in radius is a factor of six, then conduction is blocked. With the change again at 2.025 cm, the amplitude of the pulse at 2.0 cm is 20 m.v. and at 2.05 cm it is below threshold (11 m.v.).

An analytical, approximate theory for small changes in diameter has been given by Markin and Pastushenko [16].

A similar set of calculations for geometrically altered axons, sharp changes in diameter, tapers, and branches, has been carried out for a simpler set of equations by Goldstein and Rall [17]. The equations are

$$\frac{\partial^2 v}{\partial x^2} - \frac{\partial v}{\partial t} = v - \varepsilon(1-v) + u\,(v + 0.1) \tag{20}$$

$$\frac{\partial \varepsilon}{\partial t} = k_1\,v^2 + k_2\,v^4 - k_3\,\varepsilon - k_4\,u \tag{21}$$

$$\frac{\partial u}{\partial t} = k_5\,\varepsilon + k_6\,\varepsilon\,u - k_7\,u \tag{22}$$

This is a simpler set than the Hodgkin-Huxley equations but has similar properties.

c. Remarks

This sampling of the numerical results should indicate that the Hodgkin-Huxley equations provide a reasonable representation for many nerve axon phenomena. They have become a useful working tool for the experimenter. In addition the numerical solutions are helpful, with laboratory experiments, in suggesting qualitative approaches to the theory. We shall now consider such treatments.

3. QUALITATIVE RESULTS

From the computations described in section 2, it can be seen that the Hodgkin-Huxley equations yield reasonable results for many cases which are also experimentally measurable. The equations themselves and the calculations give rise to a number of mathematical questions. As we shall see, progress toward a qualitative understanding of the fifth order partial differential equation system has not been as rapid as that achieved with the related ordinary differential equation systems or with the approximation systems briefly mentioned earlier (equations (4) and (5)). However, in any case, there are a number of mathematical phenomena associated with these non-linear parabolic equations that seem to be well worth dealing with for their own sake as mathematics. It is not at all clear that these mathematically motivated investigations will reflect back on physiological knowledge, although one can be hopeful.

We will discuss the second order and third order approximation systems first and then relate what has been accomplished for Hodgkin-Huxley theory.

a. K.P.P. equations

As has been mentioned earlier, a single diffusion equation with a non-linear term is a model of several types of diffusing chemical reactions. It has also been used to describe a particular kind of genetic behavior [18]. A large number of numerical solutions have been found, most of them closely related to the chemical problems. "Non-linear eigenvalue" problems and bifurcation methods have been used in analytical approaches [19]. Of interest to us is some work of Kolmogoroff, Petrovsky, and Piscounoff [20] (we will refer to all the second order systems as K.P.P. equations regardless of the form of the non-linearity). Motivated by the equation Fisher used

$$\frac{\partial u}{\partial t} = \frac{\partial^2 u}{\partial x^2} + F(u),$$ (23)

K.P.P. showed that initial values of the form

$$u(x,0) = 0 \qquad x \le a < 0,$$
$$= g(x) \quad a \le x \le b$$ (24)
$$= 1 \qquad 0 < b \le x$$

are propagated out to an asymptotic state given by

$$u(x,t) = u(x-2t) .$$ (25)

Here u satisfies

$$u'' + 2u' + F(u) = 0$$ (26)

$$u = 0 \quad x \to -\infty$$
$$u = 1 \quad x \to +\infty$$ (27)

and F(u) has the properties:

$$F(0) = F(1) = 0$$
(i) $$\qquad F(u) > 0, \qquad 0 < u < 1$$ (28)
$$F'(0) > 0$$

F'(u) is bounded and continuous on the interval (0,1) and F(u) is at least several times differentiable.

Aronson and Weinberger [21] have generalized the K.P.P. results and also those of Kanel' [22]. In addition to the quadratic-like non-linearity described above (figure 20), they and Kanel' have shown that stable asymptotic propagating waves are attained for the following definitions of F(u):

(ii) F(0) = F(1) = 0, F'(0) > 0, F'(1) > 0, F(u) > 0 in (0,a), F(u) < 0, in (a,1) for some a in (0,1). (29)

(iii) (Kanel') F(0) = F(1) = 0, F'(0) < 0, F(u) < 0 in (0,a), F(u) > 0 in (a,1) for some a in (0,1) and for $\int_0^1 F(u) \, du > 0$ (30)

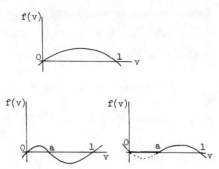

Figure 20.   Non-linear functions used to investigate propagating front solu-

tions to K.P.P. equations.

     or

     $F(0) = F(1) = 0$, $F(u) < 0$ in $(0,a)$, $F(u) > 0$ in $(a,1)$ for   a   in

$(0,1)$, $\int_0^1 F(u)\,du > 0$ (31)

     In all of these cases, there is found to be a value of $\theta = \theta^* > 0$ such

that for every initial value which lies in a bounded region and which does

propagate it propagates at the asymptotic velocity θ^*. For the special case

(i), all values of $\theta > \theta^* = 2\sqrt{F'(0)}$ propagate waves but only the value $\theta = \theta^*$

is stably attained.

     Aronson and Weinberger also consider the initial-boundary value problem

for equation (23). The initial-boundary data are

$$u(x,0) = 0, \qquad 0 \le x < \infty$$
$$u(0,t) = \psi(t) \quad 0 < t$$

(32)

Here again, the boundary value can propagate at an asymptotically stable

velocity, θ^*, corresponding to a solution $u(x,t) = u(x-\theta^* t)$ of the related

ordinary differential equation. The boundary conditions given correspond to a

physiologically stimulated axon membrane, although the equation does not.

     If one compares equation (23) with the Hodgkin-Huxley equations, the

solutions are similar to what occurs when h and n are held constant and only

v and m vary. In this case, only a rising front will propagate but it is found, numerically, that the velocity of propagation is quite close to that of the complete pulse. Such a condition, h and n constant and only a propagating front occurs experimentally when the drug tetraethylammonium chloride is added to the bathing solution of a squid axon [23].

b. The FitzHugh-Nagumo equation

The simplest non-linear diffusion equation (23) provides only a moving front. It lacks at least a second time-scale, a slow time within which the recovery portion of the pulse can be formed. Complicated ionic behaviors are not represented. To remedy this, without going all the way to the complex and empirical form of the Hodgkin-Huxley equations, FitzHugh [24] proposed another approximate system. These equations were, investigated by Nagumo, Arimoto, and Yoshizawa, [25] and the resulting system has proved to be tractable and interesting:

$$\frac{\partial^2 v}{\partial x^2} - \frac{\partial v}{\partial t} - f(v) - z = 0 \tag{33}$$

$$\frac{\partial z}{\partial t} = \varepsilon\, v - \gamma z \tag{34}$$

where f(v) is usually taken to be a cubic:

$$f(v) = v(1-v)(a-v). \tag{35}$$

FitzHugh has discussed the relationship between the space-clamped version of this system and the Hodgkin-Huxley equations from a geometrical point of view. Casten, Cohen and Lagerstrom [26] have exhibited a quantitative relationship but there is, as yet, no derivation of the FitzHugh-Nagumo equations from the Hodgkin-Huxley theory. In fact, a formal perturbation process for τ_m small, τ_n and τ_h large would probably give an approximate system somewhat different from the FitzHugh-Nagumo system.

Equation (33) by itself with z = 0, as we have seen, would produce only a moving front. It represents both the v and m equations of the more complete theory. Equation (34) represents the n and h equations.

Suppose that we consider the three boundary-value problems that we have discussed earlier for numerical computations:

(i)   Space clamped

For $\frac{\partial}{\partial x} = 0$,   the third order ordinary differential equation system is
van der Pol-like for $\gamma = 0$. For this special case and with no forcing current,
the single critical point of the system is stable and there are no oscillations.
Troy [27] has studied the oscillatory behavior for values of $\gamma > 0$, $I > 0$,
using Hopf bifurcation techniques. Troy's analysis indicates that there are
two separated regions in which periodic solutions can occur, defined by
intervals in $I > 0$ for $\gamma > 0$.

There results are useful not only for themselves but for some light they
shed on the meaning of the periodic solutions which will shortly be described
for the propagating case.

(ii) Propagation

If $v = v(x+\theta t)$, $z = z(x+\theta t)$ then the equations become

$$v'' - \theta v' - f(v) - z = 0 \tag{36}$$

$$\theta z' = \varepsilon v - \gamma z \tag{37}$$

Hastings [28], and Carpenter [29] have shown, independently, and by somewhat
different methods that pulse, or solitary wave solutions, exist. In a three
dimensional phase plane $(v',v,$ and $z)$ such solutions will be homoclinic; they
issue from a critical point and return to the same point. Hastings uses
methods involving only open sets and proves there are pulse solutions for $\gamma = 0$,
ε small. Carpenter makes use of the fact that for the limiting case $\varepsilon = 0$, the
degenerate homoclinic solution may be described by means of heteroclinic
trajectories, connecting singular points, and jump solutions (in the degenerate
solution, $\varepsilon = 0$, these are jumps of zero amplitude). This holds for $\varepsilon > 0$
and small as well, in a perturbation sense and Carpenter proves that it is
rigorously true for $\gamma \geq 0$ by using the methods of isolating blocks. These
topological methods appear to have great power in dealing with ordinary differential equations of this kind in which the behavior at singular point must be
extended to other parts of the solution space.

A non-rigorous singular perturbation analysis in powers of ε has been
carried out by Casten, Cohen, and Lagerstrom which literally patches together

the short time and long time segments of the solution for small $\varepsilon > 0$, $\gamma = 0$.
Explicit representations of the solution are given.

Periodic solutions for $\gamma = 0$ have been proven to exist by Hastings [30]
and Conley [31] and by Carpenter [29] for $\gamma \geq 0$ by the same methods as were
mentioned above. Casten, Cohen, and Lagerstrom give an explicit relationship
between the period of the solution and the wave speed but only for $\gamma = 0$. If
we recall that for $\gamma = 0$, the space-clamped version of this equation does not
have oscillatory solutions, then these periodic solutions for propagating waves
with $\gamma = 0$ are only actually achievable by forcing the system at just the
correct periodic frequency.

Hastings has reported that he has been able to prove that a second pulse
solution exists for the FitzHugh-Nagumo equations [32]. Casten, Cohen, and
Lagerstrom calculate this second traveling pulse as a perturbation from a wave
of zero speed. This is again a singular perturbation, involving an expansion
of the speed in powers of $\varepsilon^{1/2}$. Figure (21) indicates the results of the
various analyses and numerical calculation carried out by Nagumo et al. and
independently, James Cooley of IBM Research.

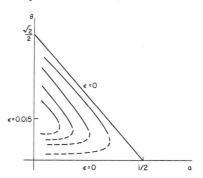

Figure 21. Speed of pulse propagation for the F.N. equations as a function
of a and ε.

Upper and lower bounds on these traveling wave velocities for solitary
pulse have been given by Sleeman and Green [33].

There is a version of this simplified, third-order system which is even
more tractable than the system we have just been discussing. McKean [34] and

Keller and Rinzel [35] have studied equations (33) and (34) with

$$f(v) = v - H(v-a), \quad 0 < a < \frac{1}{2} \qquad (38)$$

where
$$H(v-a) = 0, \qquad v \leq a$$
$$\qquad\qquad = 1 \qquad v > 0 \qquad (39)$$

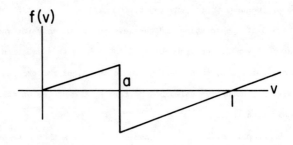

Figure 22.   The piece-wise linear function used by McKean [34] and Keller
             and Rinzel [35].

For this piecewise linear case, explicit solutions have been given for $\varepsilon > 0$
(but of any size), $\gamma = 0$. Furthermore, periodic solutions can also be given as

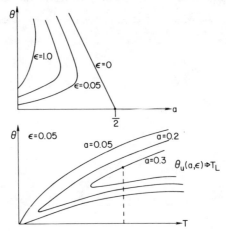

Figure 23.   Solitary pulse (upper drawing) and periodic pulse speeds as
             functions of a and ε and the period, T, for the piece-wise
             linear function of figure 22.[*]

*Reprinted from Biophysical Journal (edited by F.A. Dodge, Jr.), Vol. 13
(1973), pp. 1313-1337, with permission of Rockefeller University Press.

well as the relationships between period and pulse speed. These are indicated
in figure (23).

We will discuss the local stability problem for these equations a bit
later.

(iii) The initial-boundary value problem

Numerical calculations for the FitzHugh-Nagumo partial differential equations show that the propagating speed and shape are quickly attained as in the
Hodgkin-Huxley case. Rauch and Smoller [36] have very recently begun a mode of
analysis in which they replace the equations of lower order (3) by diffusion
equations in which the diffusion coefficient may go to zero. They are then
able to show qualitative features for the initial value problem for the partial
differential equations discussed for the K.P.P. equation has not been shown.

c. Hodgkin-Huxley equations

(i) Space clamped

Troy [37] has studied the fourth order ordinary differential equation system
that results when $\frac{\partial}{\partial x} = 0$ in (1),(2) and (3) in a manner similar to his analysis
of the space clamped FitzHugh-Nagumo equations. Troy has proved that if there
is a forcing current, I, then for some range in I there is a large recurrent
solution which corresponds to an infinite sequence of pulses. He has also found
a bifurcation of periodic orbits from the equilibrium solution as I passes
through a particular value. There are some restrictions on these results which
are typical of most of the ordinary differential equation results reported on
for the Hodgkin-Huxley system. In particular, the time scales τ_n and τ_h
must be very large compared to τ_m. For the recurrent solution result, which
corresponds to the repetitive frequency calculation of [5b] that shows an
infinite train, Troy found it necessary to hold n fixed and allow only h to
vary. The appearance of periodic and recurrent solutions has also earlier been
suggested by a numerical analysis of critical point behavior which had been
given by Cooley, Dodge, and Cohen [38].

Many of the ideas about both the Hodgkin-Huxley equations and the FitzHugh-
Nagumo equations appear in the studies of FitzHugh [24]. The phase plane analysis

of the fourth order system showed qualitatively how the lower order system (h,
n constant) with three critical points go over to the system that allows a
homoclinic solution for a single critical point.

(iii) Propagation

Carpenter [29] and Hastings [39] have studied the traveling wave solution
for the fifth order system, that is obtained when $v(x,t) = v(x+\theta t)$ in (1), (2)
and (3). In both cases, use is made of the fact that τ_n and τ_h are large
compared to τ_m. Carpenter adds the assumption that effectively, τ_m itself
is very small. Hastings' analysis requires some special assumptions, one of
which is that the ratio $\dfrac{\tau_n}{\tau_h}$ is to be restricted, which does not seem to be
physiologically meaningful. He is able to show, again by hard analysis on
open sets that there exists a traveling wave solution.*

Carpenter, using the techniques of isolating blocks and singular perturbation mentioned earlier, has shown that a solitary wave, periodic waves, and
finite wave trains exist. These may be said to correspond to the results of
the numerical experiments in that there are solutions of this type.

(iv) The partial differential equation problem and stability considerations

The counterpart of the K.P.P. attainment of propagating solutions from an
initial state has not been studied for the Hodgkin-Huxley equations. In fact,
the only analysis of the non-linear partial differential equation system itself
has been a proof of the existence of unique, bounded solutions. Evans and
Shenk [40] have given this proof together with some indications of the dependence of such solutions on initial conditions.

Other studies involve linearizations of the partial differential equations.
One may assume that a traveling wave exists and then ask whether a local,
imposed perturbation of such a solution grows or dies out. For Hodgkin-Huxley
theory, Evans [41] has set out a framework for such a calculation. In somewhat
simpler terms the problem may be described as follows:

For the equations
$$C \frac{\partial v}{\partial t} + I_i = \frac{a}{2R_i} \frac{\partial^2 v}{\partial x^2}$$

*Hastings has indicated, in private communication, that he believes he can
 obtain the second, lower amplitude, lower speed solution mentioned in section 2.

$$\frac{\partial m}{\partial t} = \frac{1}{\tau_m} (m_\infty(v) - m) \text{ etc.}$$

Let

$$v_o = v_o(x+\theta t) = v_o(\xi) \qquad (40)$$

$$m = m_o(\xi) \qquad (41)$$

etc.

be a pulse solution with velocity, θ. Then

$$v(x,t) = v_o(\xi) + v_1(x,t) + \ldots$$

$$m(x,t) = m_o(\xi) + m_1(x,t) + \ldots$$

$$n(x,t) = n_o(\xi) + n_1(x,t) + \ldots \qquad (42)$$

$$h(x,t) = h_o(\xi) + h_1(x,t) + \ldots$$

will represent a new solution perturbed by the small distribution, $v_1(x,t)$ which

satisfies

$$C \frac{\partial v_1}{\partial t} + I_1(v_1,m_1,n_1,h_1;v_o) = \frac{a}{2R_1} \frac{\partial^2 v_1}{\partial x^2} \qquad (43)$$

where

$$I_1 = \bar{g}_{na}[m_o^3 h_o v_1 + 3m_o^2 h_o (v_o-\bar{v}_{na})m_1 + m_o^3 (v_o-\bar{v}_{na})h_1] + \bar{g}_K[4n_o^3(v_o-\bar{v}_K)n_1 + n_o^4 v_1] + \bar{g}_\ell v_1 \qquad (44)$$

and

$$\frac{\partial m_1}{\partial t} = \frac{1}{\tau_m(v_o)} (\alpha(v_o) v_1 + m_1)$$

$$\frac{\partial n_1}{\partial t} = \frac{1}{\tau_n(v_o)} (\beta(v_o) v_1 + n_1) \qquad (45)$$

$$\frac{\partial h_1}{\partial t} = \frac{1}{\tau_h(v_o)} (\gamma(v_o) v_1 + h_1)$$

with

$$\alpha(v_o) = (\frac{\partial m_\infty}{\partial v})_{v=v_o} - (m_\infty(v_o)-m_o)(\frac{\partial}{\partial v} \ln \tau_m(v))_{v=v_o} \qquad (46)$$

and similar expressions for $\beta(v_o)$ and $\gamma(v_o)$. These are linear equations in v_1, m_1, h_1, and n_1 in which the coefficients depend on $v_o(\xi)$, $m_o(\xi)$ etc.

It is useful to write

$$v_1(x,t) = v_1(\xi,t)$$

$$m_1(x,t) = m_1(\xi,t) \text{ etc.} \qquad (47)$$

so that we study the perturbations in a coordinate system moving with the pulse. Then (43) becomes

$$C\left(\frac{\partial v_1}{\partial t} + \theta \frac{\partial v_1}{\partial \xi}\right) + I_1 = \frac{a}{2R_1} \frac{\partial^2 v_1}{\partial \xi^2} \qquad (48)$$

$$\frac{\partial m_1}{\partial t} + \theta \frac{\partial m_1}{\partial \xi} = \frac{1}{\tau_m(v_o)} (\alpha(v_o) v_1 + m_1) \qquad (49)$$

etc.

It is to be noted that $\dfrac{\partial v_o}{\partial \xi}$, $\dfrac{\partial m_o}{\partial \xi}$ ... etc., is a solution of this system.
This expresses the fact that a translation of the pulse, $v_o(\xi)$, is also a
solution. This somewhat complicates the stability analysis but Evans has
indicated how to handle it formally.

If we now assume that

$$v_1(\xi,t) = e^{\lambda t} \bar{v}(\xi)$$
$$m_1(\xi,t) = e^{\lambda t} \bar{m}(\xi) \qquad \text{etc.} \qquad (50)$$

where λ is a complex vector then we have

$$\frac{a}{2R_i} \bar{v}'' - C \theta \bar{v}' - I_1 = C \lambda \bar{v}_1 \qquad (51)$$

$$\theta \bar{m}' - \frac{\bar{m}}{\tau_m(v_o)} - \frac{\alpha_o(v_o)}{\tau_m(v_o)} \bar{v}_1 = - \lambda \bar{m} \qquad (52)$$

etc.

with the requirement that $\bar{v}_1, \bar{m}_1$ ... is zero at $\xi = \pm \infty$. Stability depends
on the behavior of the real part of λ.

An explicit calculation for λ and its dependence on the electrical
(C, R_i), geometrical (a), and chemical $(\bar{g}_{na}, \bar{g}_K, \text{etc.})$ characteristics of
the axon has not been made for the Hodgkin-Huxley equations. Nor has this been
done for the FitzHugh-Nagumo equations with a cubic non-linearity. Keller and
Rinzel [37] have studied the stability of solitary and periodic pulses for the
FitzHugh-Nagumo equations with the piece-wise linear function for $f(v)$.

The results for the piecewise-linear case are summarized in figure (24).
The speed curve, θ vs a, for a given value of ε is re-drawn and the dotted
portion, corresponding to the slow speed, low amplitude solitary waves, is
unstable. The knee of the curve at $\theta = \theta_c$ is the point of neutral stability.

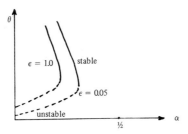

Figure 24. Stable and unstable (dotted curves) regions of the speed diagram
for the piece-wise linear non-linearity.*

For the periodic pulses, in some cases, a portion of the "fast" branch of
the speed curve is also unstable. Rinzel [42] has also studied the spatial
stability or signaling problem stability. For this case, he used coordinates
(ξ,x) rather than (ξ,t) to describe the perturbation. Further details
are given in the next paper in this volume and in [42].

d.   General remarks

As we have seen from the numerical calculations, the Hodgkin-Huxley
equations can represent a large number of the basic phenomena of pulse conduction
along nerve axons. There is one important behavior for which the original
form of the equations does not work. This is the frequency change with stimulus
change that many axons, especially in sensory systems, require. The approximate
FitzHugh-Nagumo system also lacks this characteristic.

For the Hodgkin-Huxley equations one must recognize that there is a very
strong empirical component, which until now, must be obtained experimentally.
Experimental work on the ionic permeability and conductivity change is proceeding
[4]. It is to be hoped that a formulation of these processes can be achieved
that can be included into Hodgkin-Huxley theory.

For the mathematician, the phenomena and the equations have generated
interest in non-linear parabolic systems that can behave strongly non-diffusive.

---

*See footnote to Figure 23, page 116.

Much of the qualitative work done thus far has centered on the special case of steady pulse propagation. It should be recognized that this is a kind of "intermediate asymptotic" state [43]; a stimulus must achieve threshold, there is a transient pulse build-up or attainment phase, then the steady pulse appears (actually still asymptotically), and finally at the terminus of the axon the pulse must disappear (it is not reflected).

It is not at all clear that the kind of qualitative theory that can be generated for equations of this type will be able to affect further physiological experimentation. Nevertheless, the theory does help to understand the phenomena. This is especially helpful since it appears that many other natural phenomena, because of these inherent chemical kinetic behavior, will be represented by non-linear diffusion equations.

REFERENCES

1. Herman, L., Arch. ges. Physiol., 109 (1905): 95-144.
   Bernstein, J., Arch. ges. Physiol., 92, (1902) 521-562.

2. Hodgkin, A. L. and A. F. Huxley, J. Physiol. 117 (1952) 500-544.

3. Cole, K. S., Ions, Membranes and Impulses. Berkeley, Ca.: University of California Press, 1968.

4. Hille, B. in Progress in Biophysics and Molecular Biology 21, 1970, Pergamon Press, Oxford + N.Y. pp. 1-32.

5. a. Cole, K. S., H. A. Antosiewicz and P. Rabinowitz, J. Soc. Indust. Appl. Math. 3 (1955), p. 153.

   b. FitzHugh R., and N. A. Antosiewicz, J. Soc. Indust. Appl. Math 7 (1959) p. 447.

6. Cooley, J. and F. A. Dodge, Biophys. J., 6, No. 5 (1966), 583-599.

7. Moore, J. W. and F. Ramon, J. Theo. Biol. 45 (1974) pp. 249-273.

8. Huxley, A. F., J. Physiol., 148 (1959), 80-81.

9. Dodge F. A., Intern. J. Neuroscience, 3, (1972) pp. 5-14.

10. Huxley, A. F., Ann. N.Y. Acad. Sci. $\underline{81}$ (1959) 22.

11. Mauro, A., A. Freeman, J. Cooley, and A. Cass, Biophysik $\underline{8}$ (1972),
     118-132. See also Mauro, A., F. Conti, F. Dodge, and R. Scher. J.
     gen. Physiol. (1970). 118-132.

12. Berkinblit, M. B., I. Dudzyavichus, S. A. Kovalev, S. V. Formin,
     A. V. Kholopov, and L. M. Chailakhyan, Biophysics (Biofizika) (1970),
     $\underline{15}$, 1 147-155.

13. Hodgkin, A. L. and B. Katz, J. Physiol. (1949), $\underline{109}$, 240-249.

14. Khodorov, B. I., Ye. N. Timin, S. Ya. Valenin and F. B. Gulko,
     Biophysics (Biofizika), 1970, $\underline{15}$, 1, 140-146.

15. Khodorov, B. I., Ye.N. Timin, S. Ya. Valenin, and F. B. Galko,
     Biophysics (Biofizika), 1969, $\underline{14}$, 2, 304-315.

16. Markin, V. S. and V. F. Pastushenko, Biophysics (Biofizika), 1969,
     $\underline{14}$, 316-323.

17. Goldstein, S. S. and W. Rall, Biophys. J., 14 (1974), 10, 731-758.

18. Fisher, R. Annals of Eugenics $\underline{7}$, Pt. 4 (1937), 355. See also Gazdag,
     J. and J. Canosa, J. of Appl. Prob., $\underline{11}$, (1974), 3, 445-457.

19. Cohen, D. S. (editor), Mathematical Aspects of Chemical and Biochemical
     Problems and Quantum Chemistry, $\underline{8}$, (1974), SIAM-AMS Proc. American Math.
     Soc. Providence.

20. Kolmogorov, A., I. Petrovsky, and N. Piscounov, Moscow Univ. Bull.
     Math., Serie Internationàle, Sect. A, Math. et. Mech. I, 6, 1(1937).

21. Aronson, D. G. and H. F. Weinberger, To appear in Proceedings of the
     Tulane Program in Partial Differential Equations, Lecture Notes in
     Mathematics, Springer-Verlag, 1975.

22. Kanely, I., Math. Sbornik, $\underline{59}$ (Supplement 1962), 245-288.

23. FitzHugh, R., J. Gen. Physiol. $\underline{43}$ (1960) 867-896.

24. FitzHugh, R., Biophys. J. (1961) 445-466. See also FitzHugh, R.,
     chapter in Biological Engineering, H. P. Schwan, ed., McGraw-Hill, N.Y.,
     1969.

25. Nagumo, J., S. Arimoto, and S. Yoshizawa, Proc. IRE, 50 (1962), 2061-2070.

26.   Casten, R., H. Cohen, and P. Lagerstrom, Quart. Appl Math. <u>23</u>, 4, (1975)
       365-402.

27.   Troy, W., to appear (Univ. of Pittsburgh, Pittsburgh, Pa).

28.   Hastings, S. P., to appear in Quart. J. Math.

29.   Carpenter, G., to appear (Mass Inst. of Tech., Cambridge, Mass).

30.   Hastings, S. P., Quart. J. Math. 25 (1974), 369-378.

31.   Conley, C., to appear in Proc. of the Tulane Program on partial
       differential equations, Springer-Verlag notes, 1975.

32.   Hastings, S., Personal communication.

33.   Green, M. W. and B. D. Sleeman, J. Math. Biol. <u>1</u> (1974) 153-163.

34.   McKean, H. P., Adv. in Math. <u>4</u> (1970) 209-223.

35.   Keller, J. and J. Rinzel, Biophys. J. <u>13</u> (1973). 1313-1337.

36.   Rauch, J. and J. Smoller, to appear (Univ. of Michigan, Ann Arbor, Mn.).

37.   Troy, W., to appear (Univ. of Pittsburgh, Pittsburgh, Pa).

38.   Cooley, J. F., F. A. Dodge, and H. Cohen, J. Cell and Comp. Physiol.
       <u>66</u> pp. 99-110.

39.   Hastings, S. P., to appear in Arch. Rat. Mech. Anal.

40.   Evans, J. and Shenk, N., Biophys. J. <u>10</u> (1970), 1090-1101.

41.   Evans, J., Indiana Univ. Math. J. I, <u>21</u>, (1972) p. 877, II, <u>22</u>, 1972,
       p. 75, III, 22, 1972, p. 577.

42.   Rinzel, J. to appear (N.I.H., Bethesda, Md.)

43.   Barenblatt, G. I. and Ya. B. Zeldovich, Annual Rev. of Fluid Mechanics,
       4, 1972, p. 285-312.

Lectures on Mathematics in the Life Sciences
Volume 8, 1976

SIMPLE MODEL EQUATIONS FOR ACTIVE NERVE CONDUCTION

AND PASSIVE NEURONAL INTEGRATION

John Rinzel

National Institutes of Health

Contents

I. Introduction

A primary form of communication in the nervous system is the transmission
of electrical signals by nerve cells. The signaling function of each component of a nerve cell is frequently correlated with its structure, i.e. geometrical arrangement and membrane properties. Yet, because there are many
different classes of neurons, it is not possible to provide a universal description of cellular structure and function (Shepherd [55]). However, for
some types of neurons, a classical identification applies. The cable-like
axon serves as the carrier of unattenuated nerve impulses from the cell body
to other nerve cells and muscle fibers. It transmits sequences of pulses
generated in response to inputs delivered to the cell. The input contacts on
a cell are called synapses; there is generally a large number of them (tens
of thousands). These synapses are distributed primarily over the relatively
short, but extensively branched, dendritic trees which emanate from the cell
body. The dendrites provide convergent pathways for transmission of input
signals to the cell body. Neuronal integration refers to the manner in
which synaptic input signals are transmitted and combined in the dendritic
trees and cell body.

This brief account of axonal and dendritic signal transmission is based
upon data laboriously obtained by anatomists and electrophysiologists; for
additional description see, for example, Cole [12], Hodgkin [26], and Stevens
[57]. This data along with strong biophysical intuition has led to the development of mathematical models for various aspects of cellular electrical activity. For axonal and dendritic signaling, the most familiar mathematical
descriptions can be characterized as follows.

For the conduction of nerve impulses along a uniform nerve axon, the
mathematical model involves a nonlinear parabolic partial differential equation
such as that of Hodgkin and Huxley [27]. Its solution mimics membrane potential and current as a function of time and distance along the nerve. The
typical theoretical axon is unbranched with constant diameter and often has

infinite or semi-infinite length. Thus the geometry is rather simple while
the equations are nonlinear and analytically challenging.

To model neuronal dendritic integration, the branching geometry is introduced. For certain neurons, e.g. motoneurons, the parabolic partial differential equation for membrane potential is usually assumed to be linear. It
describes passive membrane behavior. The qualitative features of its solutions are known. For an unbranched dendrite, the equation can sometimes be
solved explicitly with analytical techniques. The principle mathematical
complication is due to the extensively branched geometry.

In this chapter, I will describe simple model equations for the two modes
of cellular signal transmission outlined above. A simplified differential
equation for axonal conduction (FitzHugh [23, 24]; Nagumo, Arimoto, and
Yoshizawa [39]; and McKean [37]) and an idealized branching geometry for
neuronal integration (Rall [42]) lead to enhanced analytical tractability
without sacrifice of the important qualitative features of the phenomenology.
For axonal conduction, the traveling wave solutions corresponding to a single
impulse and repetitive trains of pulses are obtained exactly. The simple
character of the equation permits an explicit linear stability analysis of
these solutions and verification of stability conjectures made by Huxley [29]
and others [12, 14, 24, 39]. I will outline the approach and results of
these analyses and some consequences for axonal signaling by nerves. For
passive neuronal integration, the simple model has been used to qualitatively
understand the functional importance of dendritic synaptic activity. It has
also been applied quantitatively. Mathematical solutions account for experimentally recorded potentials in certain classes of neurons and provide recipes
for estimating neuronal electrical parameters. I will describe these applications in addition to analytic results which enable theoretical comparisons
of effectiveness of input delivered to individual synaptic locations.

II.   Impulse Conduction Along a Nerve Axon

1. FHN equation and a simple version

The preceding chapter [11] in this volume summarizes various aspects of
nerve impulse conduction. Therefore, I will not review that phenomenology.
As illustrated by Dr. Cohen's catalog of experimental and theoretical results,
the level of quantitative accountability of the empirical Hodgkin-Huxley (HH)
theory for squid axon is impressive. Because the equations are complicated
and nonlinear, the supporting evidence is primarily numerical. The calculations reflect but have not fully explained the underlying structure of the
mathematics of nerve conduction.

To expose that structure was FitzHugh's goal in considering a modified
relaxation oscillator as a conceptual model for nerve membrane excitability
[23]. He introduced an equation which is frequently written in the form

$$\frac{\partial v}{\partial t} = \frac{\partial^2 v}{\partial x^2} - f(v) - w \tag{1}$$

$$\frac{\partial w}{\partial t} = b\,(v - dw) \qquad , b, d \geq 0.$$

Here t is dimensionless time and x is dimensionless distance along the "nerve".
In this two-variable model, v is analogous to membrane potential and w is
referred to as a recovery variable. Without w, i.e. w=0, the nerve may reach
a stable excited level of potential from which it may not recover without an
external stimulus. The only nonlinearity in the equation is due to the function $f(v)$. It is the current-voltage law of a nonlinear conductor and corresponds to a component of membrane current. A familiar form for $f(v)$ is the
cubic function

$$f(v) = v(1-v)(a-v) \qquad , 0 < a < 1/2.$$

FitzHugh's original equation contained a cubic. A summary of his work is
given in an inspirational review article [24]. Nagumo, et al [39] also studied
the cubic case with d=0. They actually constructed an electronic transmission
line corresponding to (1); they used a tunnel diode as the nonlinear conductor.

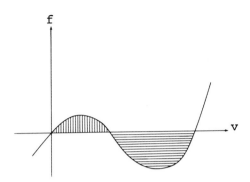

FIGURE 1. "Current-voltage characteristic", f(v) versus v, for nonlinear element of FitzHugh-Nagumo model.

For appropriate parameters, nerve like behavior such as impulse conduction were obtained for the cubic equation and the electronic nerve analog. Recent analytical results [8, 25] indicate that any f(v) with the general shape shown in Figure 1 and appropriate parameter values b and d is sufficient for impulse propagation. A typical hypothesis on f(v) is that the horizontally hatched area exceed the vertically hatched area. We will refer to equation (1) with f(v) as in Figure 1 as a FitzHugh-Nagumo (FHN) equation.

The qualitative features of the cubic FHN equation were revealed numerically and analytically by FitzHugh [23] for the case of the space clamp, i.e. v(x,t) = v(t), w(x,t) = w(t). He classified the behavior in the context of a general class of nerve conduction equations and compared it to that of the HH space-clamped equation. Mathematical aspects of repetitive firing in the space-clamped case were also considered by Troy [58] for the FHN and HH equations and by Cooley, Dodge, and Cohen [13] for the HH model. The traveling wave solutions of the cubic equation for b=d=0 have been considered by several authors [10, 24, 37, 39]. In that case of no recovery, equation (1) has a stable propagating wave front, sometimes called transition wave, solution which corresponds to successive locations along the nerve switching from the rest state to the stable excited state. Some stimulus-response or "threshold"

properties of this equation have been explored analytically [2, 59]. We note
that the FHN transition wave corresponds to a similar solution of the HH equation when sodium inactivation, h, and potassium activation, n, are held fixed
to their resting values [10]. For the cubic equation with recovery and d=0,
Casten, Cohen, and Lagerstrom [9] obtained asymptotic expressions, $0<b\ll1$,
for the solitary pulse and large wavelength periodic traveling wave solutions.

While certain features of the model have been explored analytically, the
full parametric descriptions of the solitary pulse solutions and periodic wave
trains must be carried out numerically. The same is true of the stimulus-
response properties, i.e. the functional relationship between given initial
and/or boundary data and the signal propagated by the nerve. Closely coupled
to this last point is the stability of the traveling wave solutions. Several
investigators [12, 14, 24, 29, 39] have conjectured that certain of the wave
solutions for the FHN and HH equations are unstable. It is the stable ones
which are expected to be seen in response to physiological stimuli. The
stability analysis has not been done for the wave solutions of either equation.

With this in mind, McKean [37] suggested an FHN caricature to further ease
the analytic burden. He set d=0 and chose for f(v) the piecewise linear function

$$f(v) = v - H(v-a) \qquad\qquad 0<a<1/2. \qquad (2)$$

Here H is the Heaviside step function; it satisfies

$$H(v-a) = 1 \qquad \text{for } v>a$$
$$= 0 \qquad \text{for } v\le a.$$

For this choice, the partial differential equation has piecewise constant
coefficients. In the next few sections I will outline some results obtained
for this simple model equation. Specifically, the traveling wave solutions
are found exactly and their stability is explicitly analyzed. The stability
conjectures are verified for this simple nerve conduction equation. The model
equation allows us to develop and test various stability criteria. It serves
as an instructional tool. Certain of the qualitative results formally extend

to a general class of nerve conduction equations which includes the HH and FHN
equations.

2. Traveling wave solutions

A traveling wave solution corresponds to a signal which propagates with
constant speed and shape along a uniform nerve fiber. It is mathematically
convenient since it satisfies an ordinary differential equation. As a solution to a fully specified problem for the partial differential equation it is
an idealization since it satisfies only rather special boundary and/or initial
conditions. It's relevance is as an asymptotic state which is attained for
distances away from and/or times after stimulus application. For HH squid
axon, steadily propagating signals are observed not far away from spatially
restricted stimuli; cf. calculated solutions of Cooley and Dodge [14]. I find
it interesting that the length of an experimental segment of squid axon is
typically about 8 cm. while the spatial "width" of the pulse, recorded under
normal conditions, is about 2 cm. Hence, the asymptotic waveform is realized
over a length which is only four times the pulse width.

For deducing physiological implications from mathematical analysis, we
should acknowledge however that the nonexistence of traveling wave solutions
for a given set of parameter values means only that the "nerve" does not
operate in this idealized manner. One must remember which distances and time
scales are of importance to the nerve. It may work quite well in its place
yet not admit to our neat analysis. A "graded" form of decremental signaling
has been studied numerically by Sabah and Leibovic [54] for a modified HH
equation.

A traveling wave solution of (1) is a solution of the form

$$v(x,t) = v_c(z)$$
$$w(x,t) = w_c(z) \qquad z = kx - \omega t . \qquad (3)$$

Here k and ω are positive constants for a rightward moving wave. Such a
solution must satisfy the nonlinear ordinary differential equation

$$- \omega v_c' = k^2 v_c'' - f(v_c) - w_c$$

$$- \omega w_c' = b v_c \quad . \tag{4}$$

Note, we have taken d=0 here.

For the case of a solitary pulse we seek a solution $(v_c(x-ct), w_c(x-ct))$, c>0, which tends to zero as $|z| \to \infty$, see Figure 2. It corresponds to a trajectory in the phase space (v_c, v_c', w_c) which leaves the origin as z decreases from ∞, passes through an "active" region v>a, and returns to the origin as z tends to $-\infty$. Only for select values of c is there such a solution.

I will now briefly outline the procedure for obtaining the solitary pulse solution "exactly"; for details see Rinzel and Keller [50]. This case illustrates the straight forward technique for obtaining particular solutions to the simplified FHN equation. Since equation (4) has piecewise constant coefficients, its solutions are sums of exponentials $\exp(\alpha_i z)$, i=1,2,3 where the α_i are zeros of the cubic

$$p(\alpha) = \alpha^3 + c\alpha^2 - \alpha + b/c \quad . \tag{5}$$

These zeros satisfy

$$\alpha_1 < 0; \ \text{Re}\alpha_2, \ \text{Re}\alpha_3 > 0 \quad . \tag{6}$$

This distribution of zeros reflects the fact that the unique singular point (0,0,0) in the phase space has one outgoing trajectory and a two dimensional surface of incoming trajectories for decreasing z. For the pulse shape shown in Figure 2, it follows from (6) that for z≥0 the solution must be given by $v_c(z) = a \exp(\alpha_1 z)$. By using the continuity of v_c and v_c' along with the jump condition, $v_c''(0^+) - v_c''(0^-) = 1$, we match the exponentials across z=0 and define the solution for $z_1 \leq z < 0$. Similarly, the solution is extended to $z < z_1$. Since a solitary pulse satisfies the boundary condition $v_c \to 0$ as $z \to -\infty$, the coefficient of $\exp(\alpha_1 z)$ for $z < z_1$ must be equal to zero. This boundary condition is thus equivalent to a transcendental equation which relates the pulse speed c to the parameters a and b:

$$F(a,b,c) = 0 \quad . \tag{7}$$

The pulse is determined "exactly" when its speed is found from (7).

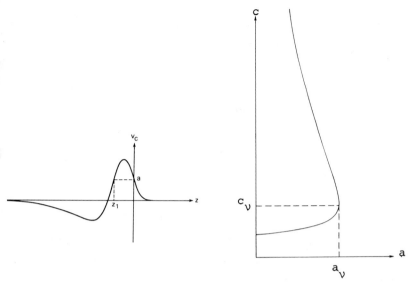

FIGURE 2. Solitary pulse traveling wave solution (left) $v_c(z)$ of equations (1) and (2) where z=x-ct. There is an associated "recovery" pulse $w_c(z)$ not shown here. A traveling pulse also satisfies equations (2) and (4) with k=1 and ω=c. Pulse speed curve (right) shows propagation speed c versus a for fixed b. Here b=.05 and the "knee" coordinates are a_ν = .35, c_ν = .43.

Rinzel and Keller [50] have solved this equation numerically and asymptotically for c in terms of a and b. For b=.05, the speed as a function of a is shown in Figure 2. For each value of a<a_ν, there are two different solitary pulse solutions with different speeds. For a=a_ν there is a unique pulse with speed c_ν, and for a>a_ν there are no solitary pulse traveling wave solutions. This speed curve has the double branched nature of parametric speed curves which have been computed for other nerve conduction equations [10, 12, 14, 24, 39]. As found for the other models, the fast pulse solution of (1)-(2) has larger amplitude than the slow one. For all the nerve conduction models, it is the slower pulse which is thought to be an unstable solution of the partial differential equation. The conjecture is made because the nerve impulse recorded from squid axon and the pulse observed on Nagumo's analog correspond to the fast solitary pulse solutions of the HH and FHN equations respectively.

Similarly, numerical integration of the partial differential equation for
physiologically reasonable initial-boundary value problems shows only the fast
pulse successfully propagating away from a source.

The steady repetitive firing of a nerve corresponds to a periodic solution
of (4). Here we seek the appropriate values of frequency ω and wave number
k for which the period of the phase z equals 2π. The reciprocal of k is proportional to the wavelength or spacing between pulse fronts in the periodic
train. The propagation speed c of the wave train is

$$c = \omega/k \qquad . \tag{8}$$

With suitable model parameters, the simplified FHN equation has a one parameter family of periodic traveling wave solutions. In the three dimensional
phase space for equation (4), a periodic solution corresponds to a closed orbit.

The periodic traveling wave solutions of the model equation are also obtained "exactly" by matching exponential solutions. Such a solution, as illustrated
in Figure 3, must satisfy the periodic boundary conditions at z=0 and z=2π. This
leads to a transcendental equation which involves k and ω, and the model parameters a and b. The solution ω to this equation for particular values of k, a,
and b defines the dispersion relation

$$\omega = \omega(k, a, b) \qquad . \tag{9}$$

Rinzel and Keller [50] obtained the dispersion relation numerically and dis-
played propagation speed as a function of wavelength for several values of
a and b. In Figure 3 (right), the dispersion curve is shown for a=.3, b=.05.
As in the case of the solitary pulse there is a nonuniqueness property. For
each $\omega < \omega_{max}$ there are two different periodic wave trains with different wave
numbers; the one with larger wave number travels slower. Hence the right
branch of the dispersion curve corresponds to the slow wave trains. For a
given wavelength, the faster solution has larger amplitude than the slower one.
As in the solitary pulse case, it is natural to conjecture that the slower
waves are unstable.

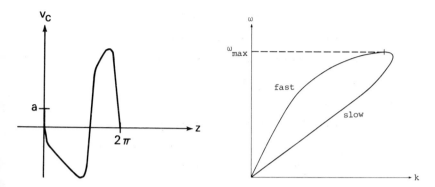

FIGURE 3. Periodic traveling wave solution (left) $v_c(z)$ of equations (1) and (2) with $z=kx-\omega t$ and period 2π. Wavelength equals $2\pi/k$ and firing frequency is $\omega/2\pi$. Dispersion curve (right) for periodic wave trains shows ω versus k for fixed a and b. Here a=.3, b=.05 and maximum frequency ω_{max} equals .129.

The one parameter family of periodic solutions corresponds to the different frequencies at which a nerve may fire repetitively. The dispersion curve shows that the propagation speed of a train depends upon its frequency. This possibility, that propagation speed depends upon firing frequency, is apparently not generally discussed by physiologists. We note that the dispersion relation has not yet been computed for any other nerve conduction equation. The importance of such dispersive effects will depend of course upon the properties of the particular nerve and its functional role. Intuitively, it is fibers which are long (in terms of pulse wavelength) and which are capable of a broad range of firing frequencies where the consequences of dispersion should be evident. The results for this simple model suggest a laboratory experiment to detect these effects. One experiment somewhat of this nature was performed by Arshavskii, Berkinblit, and Dunin-Barkovskii [3].

We also observe from the dispersion curve, that as k and ω tend to zero, the slope of the fast (slow) branch tends to the speed of the fast (slow) solitary pulse solution. Correspondingly, the solitary pulse solution is a limiting case of the periodic wave train as the wavelength becomes infinite.

This was also revealed by Casten, Cohen, and Lagerstrom [9] for the cubic
FHN equation. To understand how the solitary pulse is obtained in the limit
k→0, we consider the phase space trajectories. For k<<1, the phase point
(v_c, v_c', w_c) stays near the origin an exceedingly large percentage of the
period 2π. Analogously, for very low firing frequency, a given location on
the nerve spends more time approximately at rest. As k tends toward zero,
the closed orbit comes in contact with the origin. For k=0, the solution is
no longer periodic; the temporal period and spatial wavelength are infinite.
The trajectory which begins and ends on the singular point (0,0,0) corresponds
to a solitary pulse.

Here we have chosen to consider two types of traveling waves: periodic and
solitary pulse. For the FHN equation, analytical work indicates that these
exhaust the interesting types. For the HH equation, or a slightly modified
version, there may also be steady propagating trains with a finite number of
pulses [1, 8]. The analysis of the bounded phase space trajectories presents
interesting and challenging mathematical problems.

3. Stability of traveling waves

Because the model equation is simple and because its traveling wave solutions are known exactly, we can use it to consider various stability criteria.
Here we examine two notions of linear stability, temporal and spatial. We
describe the results for the simple model equation and indicate the implications for a general class of nerve conduction equations. For the details of
the temporal stability analysis, see Rinzel and Keller [50]; for the spatial
stability analysis, see Rinzel [51].

a. Temporal stability

One method for analyzing stability is to study the temporal evolution of
perturbations which are imposed at some time, t=0 say. This is temporal
stability analysis; it is suitable for initial value problems. For a traveling wave solution we use a moving coordinate frame z,t where $-\infty<z<\infty$ and t>0.
In these coordinates, equation (1) becomes

$$\frac{\partial v}{\partial t} = k^2 \frac{\partial^2 v}{\partial z^2} + \omega \frac{\partial v}{\partial z} - f(v) - w$$

$$\frac{\partial w}{\partial t} = \omega \frac{\partial w}{\partial z} + b\,v$$

(10)

and it has (v_c, w_c) as a t-independent solution. Temporal stability means for an initial state (t=0) "near" (v_c, w_c) that the solution to (10) tends to some translate of the wave, $(v_c(z+h), w_c(z+h))$, as t→∞. Here h is a constant and "near" implies a choice of a suitable norm.

As the basis of a stability analysis, we study the linear temporal stability of (v_c, w_c). This linear treatment does not describe the precise evolution, for all t>0, of a perturbation to the wave. Rather, it provides the indicator for whether or not the wave is stable to sufficiently small per-turbations. Various aspects of temporal stability for solitary pulse solutions to nerve conduction equations are considered by Evans [19-22]. For linear stability, we consider the linear variational equation

$$\frac{\partial V}{\partial t} = k^2 \frac{\partial^2 V}{\partial z^2} + \omega \frac{\partial V}{\partial z} - f'(v_c)\, V - W$$

$$\frac{\partial W}{\partial t} = \omega \frac{\partial W}{\partial z} + b\,V \quad .$$

(11)

We seek a solution of the form

$$V(z,t) = e^{\mu t}\, V_T(z)$$

$$W(z,t) = e^{\mu t}\, W_T(z).$$

(12)

If (11) has a solution (12) with Reμ>0 and bounded (V_T, W_T), then we say that the wave (v_c, w_c) is temporally unstable. The function (V_T, W_T) is called an unstable mode; Reμ is its growth rate. In this case, solution (12) corresponds to an infinitesimal perturbation which is bounded in x but grows with t.

For any solution to (11) of the form (12), (V_T, W_T) satisfies an ordinary differential equation which contains μ as a parameter:

$$\mu \, V_T = k^2 \, V_T'' + \omega \, V_T' - f'(v_c)V_T - W_T$$

$$\quad (13)$$

$$\mu \, W_T = \omega \, W_T' + b \, V_T \, .$$

This equation has piecewise constant coefficients; its soluticons are sums of exponentials. Note that, $f'(v_c) = 1 - \delta(v_c - a)$, where $\delta(\cdot)$ is the Dirac delta function; its presence means that V_T' undergoes a jump at values of z for which $v_c = a$. With this understanding, we seek values of μ for which (13) has a bounded solution for $-\infty < z < \infty$.

First we observe that $(V_T, \, W_T) = (v_c', \, w_c')$ is a solution with $\mu = 0$. This follows from the fact that any translate of $(v_c, \, w_c)$ is a t-independent solution of (11). Temporal stability was defined in a way to reflect this translation property.

To verify the stability conjectures we construct bounded solutions to (13) by patching exponentials. The procedure is analogous to that outlined in the preceding section. Because we require that (V_T, W_T) must satisfy certain boundary conditions, we are led to a transcendental equation which involves μ, the model parameters a and b, and the wave parameters, either k and ω for a periodic wave, or c for a solitary pulse.

For each slow solitary pulse, we have found a positive growth rate $\mu > 0$ by numerically solving such a "characteristic" equation; the associated unstable mode tends to zero as $|z| \to \infty$. In Figure 4 (left) we redraw the solitary pulse speed curve from Figure 2; the slow branch, shown dashed here, distinguishes those solutions which we have shown to be unstable. Our calculation also shows that the growth rate passes through zero and becomes negative as the speed curve knee $(a_\nu, \, c_\nu)$ is rounded. This indicates that fast pulses are stable and the unique pulse $(v_{c_\nu}, \, w_{c_\nu})$ is neutrally stable. These results for the model equation suggest a formal proof for the neutral stability of the pulse with speed c_ν for a general class of nerve conduction equations; see Rinzel [52].

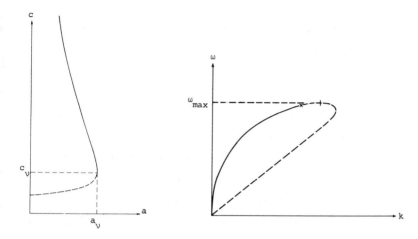

FIGURE 4. Solitary pulse speed curve (left) from Figure 2; dashed portion indicates temporally unstable pulse solutions. Neutral stability for a_ν, c_ν. Dispersion curve (right) for periodic wave trains with a=.25, b=.005; dashed portion indicates temporally unstable wave trains.

For certain of the periodic wave trains, we have solved numerically such a characteristic equation and have found positive growth rates. For this periodic case, the associated unstable mode is also periodic. We considered the dispersion curves for several values of a and b. For each one, the long wavelength (k small) slow trains were unstable. This is consistent with our results for the slow solitary pulse. As k tends to zero, the growth rate of the periodic unstable mode tends to the growth rate of the unstable mode for the slow pulse. For certain values of a and b, the entire slow branch and a portion of the fast branch was found to be unstable. Figure 4 (right) is one example of this. It illustrates the dispersion curve for b=.005 and a=.25; the dashed portion of the curve corresponds to the unstable waves.

Comment on Neutral Stability:

Our calculations did not reveal which of the periodic trains were neutrally stable. They do suggest however that neutral stability does not occur with $\mathrm{Im}\mu=0$. Whether or not this feature translates to the general class of nerve conduction equations is unknown; we see no reason to expect otherwise. Our calculations also indicate that each fast wave train is not necessarily temporally stable.

b. Spatial stability

The mathematical formulations of many neuroelectric signaling phenomena lead to boundary value problems. For example, many nerves fire repetitively in response to a time independent stimulus applied at a fixed location on the nerve. A neurophysiologist is frequently interested in the steady firing frequency as a function of stimulus strength [14, 56, 57]. A simple mathematical formulation might neglect the transient behavior, just after the stimulus is turned on, and consider only the steady repetitive firing aspect. This means solving a boundary value problem with data imposed at a fixed location, x=0 say, and specified for all time, $-\infty < t < \infty$. For such signaling problems we are interested in the transmission along the nerve of disturbances applied, for all t, at x=0. Temporal stability is not appropriate here; the coordinates z, t are not natural for such boundary value problems. Rather, we apply the concept of spatial stability.

To study stability of traveling wave solutions in this context we use the moving coordinate frame z, x where $-\infty < z < \infty$ and $x \geq 0$. In these coordinates, equation (1) becomes

$$- \omega \frac{\partial v}{\partial z} = k^2 \frac{\partial^2 v}{\partial z^2} + 2k \frac{\partial^2 v}{\partial z \partial x} + \frac{\partial^2 v}{\partial x^2} - f(v) - w$$

$$- \omega \frac{\partial w}{\partial z} = b v \quad .$$

(14)

The traveling wave (v_c, w_c) is an x-independent solution of (14).

Spatial stability means that for a signal which is imposed at x=0 and which is "near" (v_c, w_c) the solution to (14) tends to a translate of (v_c, w_c) as $x \to \infty$. For example, think of the simplest periodic case. Suppose the signal imposed at x=0 is $(v_c(-\omega t), w_c(-\omega t))$ with a small superimposed disturbance of period $2\pi/\omega$. If the wave is spatially stable, the perturbation will decay with distance from the source and the wave will be successfully transmitted to distant locations along the nerve.

As a first step in analyzing the spatial stability of the traveling wave solutions, we study their linear stability. To do this, we consider the linear variational equation

$$- \omega \frac{\partial V}{\partial z} = k^2 \frac{\partial^2 V}{\partial z^2} + 2k \frac{\partial^2 V}{\partial z \partial x} + \frac{\partial^2 V}{\partial x^2} - f'(v_c) V - W$$

(15)

$$- \omega \frac{\partial W}{\partial z} = b V \quad .$$

We now look for solutions to (15) of the form

$$V(z, x) = e^{\lambda x} V_S(z)$$

(16)

$$W(z, x) = e^{\lambda x} W_S(z) \quad .$$

It follows that (V_S, W_S) must satisfy the ordinary differential equation

$$- \omega V_S' = k^2 V_S'' + 2k\lambda V_S' + \lambda^2 V_S - f'(v_c) V_S - W_S$$

(17)

$$- \omega W_S' = b V_S \quad .$$

The traveling wave (v_c, w_c) is spatially unstable if for any λ with $\text{Re}\lambda > 0$, equation (17) has a bounded solution which belongs to an admissible class of perturbations. The function (V_S, W_S) is an unstable mode and $\text{Re}\lambda$ is its spatial growth parameter. The solution (16) to (15) corresponds to an infinitesimal perturbation which is bounded for all t, $-\infty < t < \infty$, but grows with x. Here, as in the temporal stability case, (v_c', w_c') is a solution to (17) with $\lambda = 0$.

By patching exponentials, Rinzel [51] has constructed unstable modes for the slow traveling wave solutions of the simple model equation for various values of a and b. The growth rates were obtained by solving a transcendental equation. The results again confirm the stability conjectures. Each slow solitary pulse is spatially unstable; the unique pulse (v_{c_ν}, w_{c_ν}) has neutral spatial stability. The stability picture Figure 5 (left) is the same as in Figure 4 (left). For the periodic case, the slow wave trains were found to be unstable; the growth rates were determined for periodic unstable modes. The calculations further revealed that the wave train with maximum frequency is neutrally stable. That is, the growth rate passes through zero and becomes negative as one proceeds from the slow branch to the fast branch of the dispersion curve.

This neutral stability characterization has been formally extended to a general class of nerve conduction equations (Rinzel [52]); the results for the simple model equation motivate the extension. We note that this neutral stability statement does not apply for temporal stability; the particular example of Figure 4 demonstrates that the maximum frequency wave train can be temporally unstable. To summarize our spatial stability results, the dispersion curve for b=.05, a=.3 from Figure 3 is reproduced below in Figure 5 (right); the unstable branch is shown dashed.

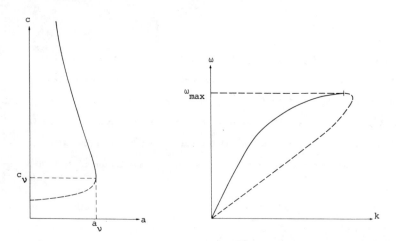

FIGURE 5. Solitary pulse speed curve (left) from Figure 2; dashed portion indicates spatially unstable pulse solutions. Neutral stability for a_ν, c_ν. Dispersion curve (right) for periodic wave trains from Figure 3; dashed portion indicates spatially unstable wave trains. Neutral stability for ω_{max}.

Remark:

From our explicit analysis emerges a qualitative difference in the characterization of neutral stability for temporal and spatial stability. It is
reasonable to expect that this difference should carry over to the general
class of nerve conduction equations. The quantitative consequences will of
course depend upon the particular equations. The moral here is that the
stability analysis should reflect the context of the biophysical setting.

In deducing functional implications from our stability results, we must
be modest. Our intuition suggests that spatially unstable wave trains will
not be observed in response to spatially localized stimuli. We acknowledge,
however, that our analysis is a linear one. To quantitatively describe the
dependence of firing frequency on stimulus properties for a particular model
requires numerical calculations; this has been done in a few cases [14], [55],
(Rinzel, unpublished). The expression of such a stimulus response result in
terms of the associated dispersion curve and spatial stability of the wave
trains is yet to be provided.

III. Neuronal Dendritic Integration

1. A mathematical model

Integration here refers to the way membrane potential is developed
throughout the cell body and dendritic trees in response to the inputs they
receive. It is a spatio-temporal process. For synaptic inputs, physiologists
have identified two types. An excitatory synaptic input produces a brief
positive deviation of membrane potential from rest, an EPSP (excitatory post
synaptic potential). An inhibitory one results in a negative deviation, an
IPSP. These potentials are the result of membrane permeability changes at
the synapse. For a given permeability change, the amplitude of an EPSP at a
single active synapse (a "unitary" event) will depend upon its location; for
example, see [34]. For a motoneuron somatic synapse, the "unitary" EPSP is
on the order of 0.1-1.0 mV [30, 31]. With sufficient excitation, the cell
will elicit an action potential near the base of its axon. The action

potential is then transmitted along the axon. A precise definition of the

implied threshold is not given. In some cases, it is thought to be a critical

amount of charge delivered to the axon [5] while in other cases a critical

value of instantaneous membrane potential is identified [17]. We will not

include here a mathematical description for action potential generation.

Our model equation will be for the transmission of potentials in the

dendrites. For certain kinds of cells, such as motoneurons, the dendritic

membrane is generally believed to be "non-active" or to have a "high thresh-

old". That is, the potential range over which membrane current depends lin-

early on membrane potential is assumed to be substantially larger than that

for axons. Synaptic potentials are said to spread passively in the dendrites.

This means that the EPSP amplitude generated at a synapse is transmitted to

other dendritic locations in an attenuated manner. The peak amplitude, i.e.

maximum over time at a given location, decreases with distance from the input

site. As a consequence of these suppositions, the mathematical model is

linear; the membrane conductance is assumed independent of potential.

We note that these assumptions are not based upon direct observation of

dendritic potentials, but rather upon recordings taken at the cell body. For

most cells, physiologists cannot provide intracellular electrical data from

dendrites; the branches are too small to penetrate consistently.

One task for the model, which is based on such assumptions, is to account

for this data from the soma. It has satisfactorily done this for a number

of investigators [4, 5, 30, 31, 36, 44]. Moreover, the model theoretically

demonstrates the functional significance of dendritic synaptic activity,

including that which originates at distal (outlying) dendritic locations

from the cell body. It is concluded that dendritic cable properties play a

major role in determining neuronal integrative behavior [5, 41, 43]. This

is consistent with the anatomical observation that most synaptic contacts are

distributed over the dendrites as opposed to the soma. While such conclusions

are now widely accepted, they offered a challenge to an earlier hypothesis that

distal dendritic inputs were functionally insignificant [17, 18]. This earlier
point of view was based upon membrane parameter estimates which were obtained
by underestimating dendritic cable properties [15]. The subsequent mathematical
modeling of Rall and others can claim significant credit for contributing insight on this issue.

The usual mathematical description of dendritic integration assumes that
each dendritic branch segment can be represented as a one dimensional core
conductor with passive membrane. The equation for the deviation of membrane
potential from rest, $v(x,t)$, in a constant diameter segment is (see, e.g.
[35] for a derivation)

$$\lambda^2 \; \frac{\partial^2 v}{\partial x^2} \; = \; \tau \; \frac{\partial v}{\partial t} \; + \; v \; + \; i_{in}(x,t) \quad . \tag{18}$$

Here λ is the electrotonic length constant for the dendritic branch; it is
proportional to the square root of the branch diameter d. The membrane time
constant is τ and it is independent of d. Successive terms on the right hand
side of this equation represent a decomposition of the membrane current density
into a capacitive component, a resting conductance component, and an input component. The input current density $i_{in}(x,t)$ accounts for synaptic input and
externally applied current. For example, when excitatory input is modeled
as a conductance change $\varepsilon(x,t)$ we would have

$$i_{in}(x,t) \; = \; \varepsilon(x,t) \; (v - v_e) \tag{19}$$

where v_e is the excitatory equilibrium potential, a constant.

Note, equation (18) is not directly derived from an active nonlinear model
such as HH or FHN by linearizing around the rest state. For such a derivation,
the non-synaptically induced, or intrinsic, ionic permeabilities of the membrane would still depend on time. Here, those permeabilities are taken to be
constant. Without synaptic input, equation (18) is often referred to, for
historical reasons, as the "cable equation" for "electrotonic potential".
Electrotonic theory has also been applied to describe small deviations from
rest in axons [28, 35].

A mathematical model for dendritic integration would include the numerous branches in the dendritic trees. For some models, the soma is approximated by a uniform membrane patch [30, 31, 41]. For others, some initial segment of each dendritic trunk accounts for the cell body [46, 47]. We follow the latter approach. Since we do not incorporate axonal signaling here, it is reasonable to neglect the current flowing into the axon.

To determine the potential distribution throughout a dendritic neuron model we must fully specify a mathematical problem. With given synaptic and/or externally applied input, a solution must satisfy, in each branch, the equation (18) in addition to certain initial and boundary conditions. It is reasonable to assume that no current flows out the tiny dendritic terminals, i.e. their distal tips. This insulated end boundary condition is $\partial v/\partial x=0$. At a dendritic branching point, the membrane potential is continuous and axial current is conserved. This latter requirement is a matching condition for the derivatives $\partial v/\partial x$ in the various segments which meet at the branching point.

In general, the problem which we have formulated is not analytically solvable. However, for a certain class of dendritic trees, the matching conditions at the branching points can be easily satisfied. Rall [42] realized this and exploited it to greatly simplify the mathematics and expose the qualitative and quantitative features of the phenomenology. In the next section we will describe Rall's simplification.

Approaches which treat arbitrary branching geometries generally require extensive computing. Recently, Barrett and Crill [4, 5] used numerical transform methods to ambitiously calculate solutions corresponding to input at single locations. Their dendritic configuration was a reconstruction of a cat spinal motoneuron from which they recorded experimentally. An impressive color picture of one of their motoneurons appears on the cover of IBM Jour. of Res. and Devel., July, 1973. Transform solutions have also been obtained by Butz and Cowan [7] for arbitrary branching geometries. Time domain expressions for their solutions generally require numerical inversion.

2. Idealized branching

It is convenient to introduce a dimensionless time variable $T=t/\tau$ and a dimensionless distance variable $X(x)$. Electrotonic distance X is measured relative to distance from the "soma", the junction of all the dendritic trunks. For a given dendritic location, the physical distance x to the soma is the sum of the distance increments Δx_j for the different branches in the path to the soma. The electrotonic distance $X(x)$ is the sum of the Δx_j each divided by the electrotonic length constant λ_j for that branch [42]:

$$X(x) = \int_0^x \lambda^{-1}(s)\,ds \quad . \tag{20}$$

Thus in a given branch

$$dX/dx = 1/\lambda \tag{21}$$

and this quantity is proportional to $1/\sqrt{d}$. Here, we will use upper case notation to express quantities in terms of X and T. The cable equation (18) is then written as

$$\frac{\partial^2 V}{\partial X^2} = \frac{\partial V}{\partial T} + V + I_{in}\,(X,T) \quad . \tag{22}$$

To motivate the idealized branching criterion we consider the simple tree illustrated in Figure 6. Here input is applied to the parent branch at $X=0$. The n identical siblings which emanate from the parent at $X=X_1$ are of length $L-X_1$. By symmetry, membrane potential is distributed identically in each of the siblings. At the branching point X_1, the axial current entering from the

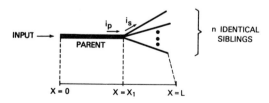

FIGURE 6. Simple configuration used to illustrate consequences of idealized dendritic branching. From a parent branch emerge n identical sibling branches. Lower scale indicates electrotonic distance.

parent i_p is equal to n times that which is leaving in each sibling i_s; $i_p = n \ i_s$.
According to cable theory, the axial current in a branch is obtained by Ohm's
law and is therefore proportional to the voltage gradient $\partial v/\partial x$ times d^2. This
last factor reflects the fact that axial resistance per unit length is inversely
proportional to cross sectional area. In terms of electrotonic distance, the
matching condition at X_1, is expressed as

$$\left.\frac{\partial V}{\partial X}\right|_{X=X_1^-} = n \left(\frac{d_s}{d_p}\right)^{3/2} \left.\frac{\partial V}{\partial X}\right|_{X=X_1^+}$$

In general, this means that $\partial V/\partial X$ will experience a jump discontinuity at
X_1. If however the factor in parentheses is equal to $1/n$, the derivative of
V is continuous at X_1. In this case, the solution to the simple problem is thus
obtained by solving (22) for $0<X<L$ with $I_{in}=0$; explicit boundary conditions are
imposed only at $X=0$ and $X=L$. Thus for branching which satisfies the criterion

$$n \ d_s^{3/2} = d_p^{3/2} \tag{23}$$

and for electrotonically symmetric input configurations, matching at the
branching points is trivially accomplished.

The fundamental notion of this simple example was described by Rall [42]
for more general branching patterns in dendritic trees. For a class of problems, the branching geometry of a dendritic tree mathematically reduces to
a single unbranched "equivalent cylinder". The principle assumptions which
allow this reduction are:

(i)   for each X, the sum of the branch diameters each raised to the 3/2-
power is equal to $d_o^{3/2}$, where d_o is the trunk diameter;

(ii) the dendritic terminals are equidistant, in terms of X, from the
soma;

(iii) inputs, initial conditions, and boundary conditions are "electrotonically" symmetric.

Anatomical measurements reported by some investigators [36, 43] indicate that the 3/2-branching criterion is satisfied to a first approximation in dendritic trees of cat motoneurons. And while other investigators [4] have found that $\sum_j d_j^{3/2}$ (X) decreases with distance from the soma, the decrease is attributed largely to tapering of dendritic branches. In this regard, Rall's theory [42] can accommodate some forms of taper.

A Sample Problem:

Consider the neuron model in Figure 7 (left). There are six identical trees each of length L and with two orders of 3/2-branching. For this case of symmetric bifurcations, branch diameter decreases by about 37% at each successive generation. A sample problem is: determine the response to a transient synaptic bombardment, a conductance change $E(T)$, which is restricted to all dendritic locations in the distance interval (L/3, 2L/3) and which is uniform in that interval. Suppose the neuron is initially at rest and its terminals are insulated. By symmetry we need consider only one of the trees. For this problem, we may apply the equivalent cylinder reduction; see the right side of Figure 7. The transient response is obtained by solving equation (22) for a single cylinder, 0<X<L, and for T>0 with the input

NEURON MODEL

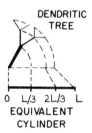

DENDRITIC
TREE

0   L/3 2L/3 L
EQUIVALENT
CYLINDER

FIGURE 7. Idealized neuron model (left) composed of six identical dendritic trees with two orders of branching. The point of common origin is regarded as the neuron soma. On the right, a dendritic tree is related to its unbranched equivalent cylinder. For the sample problem (see text) each branch length is L/3.

$$I_{in}(X,T) \;=\; E(T)\;(V-V_e) \qquad \text{for } L/3<X<2L/3$$

$$\qquad\qquad\;=\; 0 \qquad\qquad\qquad \text{otherwise,}$$

subject to the initial condition $V(X,\,0)=0$, and boundary conditions

$$\frac{\partial V}{\partial X}(O,T) \;=\; 0\;, \qquad\qquad \frac{\partial V}{\partial X}(L,T) \;=\; 0. \qquad\qquad (24)$$

The boundary condition at X=0 results from symmetry. Note that for Figure 7, the input function for the equivalent cylinder corresponds to one-half of that input applied to each dendritic branch location in the synaptically excited zone. It is important to realize that, because the equivalent cylinder concept involves a class of trees, the solution to this sample problem has application beyond the configuration in Figure 7.

3. Applications of a dendritic neuron model

a. EPSP shapes

Calculated and analytical solutions to problems similar to the sample one above illustrate several qualitative features of dendritic integration in the neuron model. They show how the theoretical soma potential depends upon input time course and location. Generally, for a given time course, more distant inputs result in more slowly-varying, broader temporal waveforms at the soma. When a given input is applied successively to different locations, the spatio-temporal pattern could determine whether or not a cell might respond with an action potential. As an example, Rall [43] compared a centrifugal pattern, roughly of the form $E(\alpha X-T)$, with a centripetal presentation $E(\alpha(L-X)+T)$ for $\alpha=1.6$; the latter resulted in a substantially larger amplitude soma potential. Similar calculations illustrated the effects of inhibition, spatial summation, and temporal summation.

Calculated EPSP's have also been quantitatively described with shape indices such as time-to-peak and half-width [32, 44]. The range of computed shape indices has accounted for experimental ones obtained from cat motoneurons in various laboratories [30, 31, 38, 44]. The associated range of input locations implicates dendritic activity which is widely distributed over the motoneuron surface and not merely restricted to soma locations. The comparisons

also reveal that significant soma depolarization could result even for inputs

identified to originate at dendritic locations of considerable distance from

the soma.

b. Neuronal parameter estimates

To determine estimates for electrical and geometrical parameters, e.g. τ

and L, of a dendritic neuron we might perform the following experiment. Inject

a brief current into the cell body with a microelectrode. After the current

is turned off, we will have established a potential distribution in the cell.

Take it as an initial condition $V_o(X)$. Suppose there are no other significant

inputs to the cell. To describe the decay back to rest of the potential dis-

tribution throughout the cell we could employ the equivalent cylinder concept.

This means solving (22) with $I_{in}=0$ subject to the boundary conditions (24) and

the initial condition $V(X, 0)=V_o(X)$. This problem can be solved explicitly

[46], e.g. by separation of variables and eigenfunction expansions. The solu-

tion is

$$V(X,T) = \sum_{n=0}^{\infty} c_n \cos(n\pi X/L) \exp[-(1+(\frac{n\pi}{L})^2)T] \qquad (25)$$

where the c_n are determined by $V_o(X)$.

From the representation (25), the large time behavior of the potential

recorded at the soma is

$$V(0,T) \sim c_1\exp(-T) + c_2\exp[-(1+\frac{\pi^2}{L^2})T], \qquad T\gg1 \quad . \qquad (26)$$

Since $T=t/\tau$, an estimate for τ can be obtained from the ultimate decay of the

membrane potential. For motoneurons, estimated values of τ are about 5 msec

[31, 36]. Occasionally, the electrophysiological data will permit one to

estimate L based on the decay rate of $V(0,T)-c_1\exp(-T)$. Estimated values

for L lie in the range of 1-2 [6, 36, 40]. Detailed anatomical measurements

[4, 36] agree with this range.

The above estimation technique is due to Rall [46]. It would also apply

of course to the decay of $V(X,T)$ after a brief burst of synaptic activity.

Techniques based upon other analytical representations of the solution for brief current injection have been described by Jack and Redman [32, 33]. Their recipes when applied to data from cat motoneurons yield estimates similar to those reported here.

c. Response for input to single branch locations

   The calculations of theoretical EPSP's which were described above were performed by treating a dendritic tree as a single unbranched equivalent cylinder. Input to only a single branch location was not explicitly considered. This latter case however can be treated analytically under the 3/2-branching assumption. The transient response $K(X,T;X_{in})$ throughout the dendritic neuron model to a pulse of current (δ-function) at a single branch site X_{in} was determined explicitly by Rinzel and Rall [49]. Their work is summarized by Rinzel [53]. They obtained their results by using Laplace transform methods and by exploiting the principle of superposition in this linear system. With this fundamental solution, one can determine the response to an arbitrary current injection $I_{in}(T)$ at X_{in} according to the convolution formula

$$V(X,T) = \int_0^T K(X,T-s;X_{in})\ I_{in}(s)\ ds \qquad . \qquad (27)$$

In (27), the initial state is taken to be $V(X,0)=0$.

   As an example, Rinzel and Rall calculated the response at various locations in the neuron model for a brief excitatory input current to a single branch terminal, $X_{in} = L$. The results are illustrated in Figure 8. The neuron model is shown upper right. There are six trees with three orders of symmetric branching. Here L=1.0 and the branch lengths are identical. The input time course, shown upper left, corresponds to an EPSP shape in the range of those seen experimentally. Normalized voltage transients are plotted on a semi-log scale for BI(branch input), its ancestors: P(parent), GP, GGP, and the soma. These dendritic transients illustrate the attenuation and broadening characteristics of passive membrane behavior. The ratio of the peak amplitude at the input terminal to the attenuated peak response at the soma is about 235 for this case.

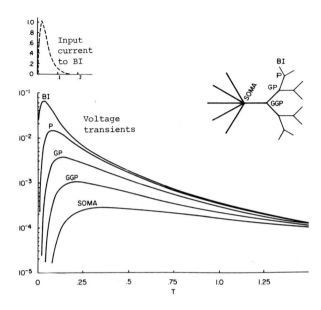

FIGURE 8. Semi-log plots of scaled membrane potential versus T at several
locations in the neuron model for transient current injected into the terminal
of one branch. The normalized input current is shown upper left. BI desig-
nates the input branch terminal while P, GP, and GGP designate the parent, grandparent, and great grandparent modes, respectively, along the mainline from BI
to the soma. The neuron model shown upper right has six trees with three orders
of branching, L=1.0, and ΔX=.25 (branch length).

Such transient solutions can also be used to compare, for different X_{in},
potentials at the input sites and at the soma. For example, suppose the same
input current as in Figure 8 is applied to the soma. The peak amplitude at
the soma in this case is about seven times that of Figure 8 for input to a
terminal. We see that, although the former case exhibited severe attenuation
to the soma, the soma peak for the distal input is not orders of magnitude
insignificant relative to that for the somatic input. The reason is that,
because of different input impedances, the amplitude at a branch input terminal
is large relative to that at a soma input location (see also, [5, 48]).
Theoretical effectiveness comparisons have also been made on the basis of charge

delivered to the soma [5, 30, 49].

We note that the difference in peak amplitudes at the soma for input at
the soma versus the dendritic location contrasts to the relative uniformity of
experimental EPSP peaks found in certain studies on cat motoneurons. If
synapses at all locations are assumed to operate in a similar manner, then one
way to explain the uniformity is to suppose that an EPSP of dendritic origin
X_{in} corresponds to several active synapses at X_{in} and a soma input corresponds
to a smaller number. This observation led Iansek and Redman [30] to a calculation of five to ten as the relative number of active synapses for dendriti-
cally impinging axons compared to those axons which contact the soma.

As a special case, Rall and Rinzel [47] applied their results to the time-
independent case of steady current injected to a single dendritic branch. This
application produced analytical expressions for the steady attenuation of
potential from X_{in} to the cell body and for the input resistance R_{in} at X_{in}.
These results show for the neuron model how R_{in} depends on the number of dendritic trees and the number of orders of branching for a given X_{in}.

For such time independent problems, I can briefly outline how we applied
superposition methods to obtain solutions; for details, see [47]. The transform solutions for the transient case are obtained in the same way. Consider
the neuron model in Figure 7. It has six trees with two orders of symmetric
bifurcations. The branch lengths are all L/3. Suppose steady current I is
applied to the terminal BI; the other terminals are insulated. We obtain the
solution as a sum of simpler component solutions as schmetized in Figure 9.
In A-C the trees are shown as their equivalent cylinders; branching is not
treated explicitly at this stage. Input to just one tree, as shown in C,
corresponds to the combination of A, I/6 to each of the trees, with B, 5I/6
to the input tree and -I/6 to each of the others. To now treat the branching,
we proceed from C; only the branching in the input tree matters. The configuration in C is redrawn in D with the input I equally divided among the branch
terminals; the second order branches are still retained as their equivalent

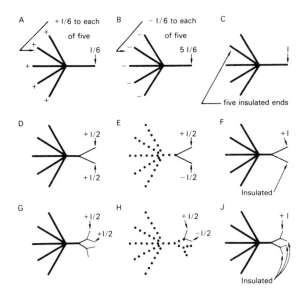

FIGURE 9. Schematic diagrams which illustrate superposition technique to obtain time independent solution for steady current applied to one branch terminal. A-C are for input to all terminals of just one tree. D-H and J consider branching details of input tree.

cylinders. The superposition of D with E leads to F. Note that symmetry implies zero potential for dendritic regions which are shown dashed. The second stage of branching is handled similarly. First, G is redrawn in F. The superposition of H with G yields J, the configuration we seek. The solution, for the input branch, is the sum of solutions to the component problems A, B, E, and H:

$$V(X,T) = (R_{T_\infty} I) \quad \frac{1}{6} \frac{\cosh X}{\sinh L} + \frac{5}{6} \frac{\sinh X}{\cosh L} + \frac{\sinh (X-L/3)}{\cosh (2L/3)} + 2 \frac{\sinh (X-2L/3)}{\cosh (L/3)}$$

Here, R_{T_∞} is a constant such that the input resistance of the neuron, seen at the soma, is given by $R_{T_\infty} \coth L/6$.

A by-product of this superposition method is an explicit demonstration that the response at the soma is independent of the way I_{in} is distributed

156   J. Rinzel

among the branch locations at X_{in}.

d. Neuron models with dendritic integration and axonal signaling

Finally I would like to briefly mention two applications in which a
neuron model was extended to include a short segment of axon with active membrane properties.

Dodge and Cooley [16] formulated a computational model of a dendritic
motoneuron with axon. By matching voltage clamp data from motoneurons, they
adapted the HH description of "excitable" (nonlinear) ionic current flows to
model membrane behavior in the active regions. For motoneurons, the axon is
myelinated. Only at discrete nodes is membrane exposed. Between the nodes,
a myelin sheath insulates the axon; here, the linear cable equation description
is appropriate. A schematic representation of their motoneuron and theoretical
cable model is given in Figure 10. The dendritic structure is represented as an

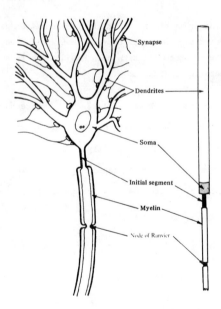

FIGURE 10. Schematic drawing of a spinal motoneuron, showing a small fraction
of all synaptic endings, and a representation of its electrical characteristics
by a nonuniform cable. (From Dodge and Cooley [16].) Reprinted with permission
of the publisher. Copyright 1973 by International Business Machines Corporation.

equivalent cylinder. Darker shading in the schematic cable indicates more highly
excitable membrane behavior. With their model they could account for various
experimental data which included action potential recordings from the soma. They
concluded that a satisfactory match with the data required passive dendritic
membrane. Further they conclude, in support of experimental interpretation
[17, 18] that soma membrane has, in addition to lower sodium conductance
density, an intrinsically "higher threshold" than the initial segment of
the axon. Based upon the agreement of the theoretical and experimental
evidence for non-uniform distribution of excitability characteristics over
the neuronal membrane, they suggest a mechanism for its development and
maintenance.

In a different application, Rall and Shepherd [45] incorporated passive
and active membrane behavior in neuron models which they used to study the
interactions between two populations of cells. Their motivation was to account
for extracellular recordings obtained in the olfactory bulb of the rabbit.
Corresponding to the experimental setup, they assumed that the two cell populations were distributed symmetrically and uniformly in a spherical arrangement. One population received direct synchronous input from action potentials
which were excited by an external stimulus at some location on their axons
and which then propagated toward the cell bodies and dendritic trees (this is
called antidromic stimulation). The symmetry and synchrony permitted the
modelers to study a representative portion of the spherical arrangement and
thereby reduce the computational effort. To calculate dendritic potentials,
they used the equivalent cylinder reduction. Their mathematical model for
active membrane behavior was different than either the HH or conceptual FHN
description. From membrane potential distributions they determined theoretical
extracellular potentials. To adequately match their calculated extracellular
potentials with the experimental ones, Rall and Shepherd postulated that the
two populations interacted by means of synaptic contacts between their respec-
tive dendrites. The notions of dendro-dendritic synapses and intercommunication
are not included in the classical structure-function description of most

158   J. Rinzel

physiologists. It is a tribute to the modeling here that, in the olfactory
bulb and other networks, anatomists subsequently discovered such connections.

IV. Summary

We have focused on two aspects of neuroelectric signaling at the cellular
level and have discussed simple mathematical descriptions for them. We adopt
a classical, but not universal, structure-function relationship for neurons.
An axon is for active nerve impulse conduction over long distances. The
equation for membrane potential is nonlinear; the geometry is simple. A
neuron's dendrites transmit input signals over relatively short distances
to the cell body. For this, passive membrane behavior is adequate in some
cells. The equation for membrane potential is linear; the geometry includes
dendritic branching.

Signal transmission over long distances along axons is mathematically
reflected by the existence of traveling wave solutions to nerve conduction
equations. We describe properties of these solutions for a model equation
which is a simplified version of the conceptual one studied by FitzHugh and
Nagumo, et al. As do other equations, this simple one has solitary pulse and
periodic traveling wave solutions. The latter correspond to steady repetitive
firing of a nerve. For this equation, these solutions are determined explicitly.
Typically, there are two solitary pulse solutions, one fast and one slow. In
the periodic case, there is a one parameter family of waves with a dispersion
relation between firing frequency and pulse spacing. For a given frequency,
there are typically two different solutions with different propagation speeds.
The stability of these traveling wave solutions is explicitly analyzed. We
consider two different notions of stability: temporal stability for initial
value problems, and spatial stability for boundary value problems. We find,
in agreement with conjectures made for solutions of other equations, that the
slow waves are unstable.

For dendritic integration, we formulate a mathematical model based upon
a linear cable equation for membrane potential in each dendritic branch. We

describe how Rall simplifies the general formulation by treating a dendritic
tree of a certain class as a single unbranched "equivalent cylinder". The
reduction requires certain symmetries and idealized branching. Various applications of this model reveal qualitative and quantitative features of dendritic membrane potential transients. For synaptic activity, the response at
the soma reflects the spatio-temporal pattern of that activity. The temporal
decay of the soma response after a brief input can be used to estimate neuronal
parameters. For trees of the equivalent cylinder class, analytic solutions
are obtained for input to just one branch site. These solutions have been
used to illustrate attenuation from branch input sites to the soma, to compare
the effectiveness of inputs at different sites, and to get expressions for
input resistance at branch sites. For motoneurons, the model has been used to
account for various experimental data, EPSP and action potential shapes, and
to implicate the functional importance of dendritic synaptic activity. Use of
the model to interpret data obtained from a rabbit's olfactory system, led to
the understanding of an expanded role for dendritic signaling in this system.

   It is worthwhile to reemphasize the approach here. The dissection of
a simple model provides us a means to formulate appropriate theoretical concepts and test our intuition. By exploiting the simplicity, it is possible
to expose certain structural features of the models. With an understanding
of them, we hope to characterize, for example, a model's capability for various
modes of signal transmission and to characterize the fundamental features of
stimuli ("thresholds") which correspond to those different modes. These
characterizations should take the form of analytic prescriptions: functional
inequalities, asymptotic expressions, etc. For a model which is quantitatively
descriptive, these prescriptions may translate into functional implications
and into formulae for identifying physiological parameters and determining
critical values of them. For a model which is more conceptual, the characterizations may suggest appropriate physiological translations for a realistic
model which shares qualitative features with the simple one.

160   J. Rinzel

Acknowledgements

   I am grateful to H. Cohen and W. Rall for helpful discussions and comments
on this manuscript.

References

1. W. J. Adelman and R. FitzHugh, "Solutions of the Hodgkin-Huxley equations
    modified for potassium accumulation in a periaxonal space", Federation Proc.
    34 (1975) 1322-1329.

2. D. G. Aronson and H. F. Weinberger, "Nonlinear diffusion in population
    genetics, combustion, and nerve propagation", in Proceedings of the
    Tulane Program in Partial Differential Equations (Lecture Notes in
    Mathematics). Berlin-Heidelberg-New York: Springer. To appear 1975.

3. Y. I. Arshavskii, M. R. Berkinblit, and V. L. Dunin-Barkovskii, "Propa-
    gation of pulses in a ring of excitable tissues", Biophysics 10 (1965)
    1160-1166.

4. J. N. Barrett and W. E. Crill, "Specific membrane properties of cat
    motoneurones", J. Physiol. (Lond.) 239 (1974) 301-324.

5. J. N. Barrett and W. E. Crill, "Influence of dendritic location and mem-
    brane properties on the effectiveness of synapses on cat motoneurones",
    J. Physiol. (Lond.) 239 (1974) 325-345.

6. R. E. Burke and G. tenBruggencate, "Electrotonic characteristics of alpha
    motoneurones of varying size", J. Physiol. (Lond.) 212 (1971) 1-10.

7. E. G. Butz and J. D. Cowan, "Transient potentials in dendritic systems of
    arbitrary geometry", Biophys. J. 14 (1974) 661-689.

8. G. A. Carpenter, "Traveling wave solutions of nerve impulse equations",
    Doctoral thesis, University of Wisconsin. 1974.

9. R. H. Casten, H. Cohen, and P. Lagerstrom, "Perturbation analysis of an
    approximation to Hodgkin-Huxley theory", Quart. Appl. Math. 32 (1975)
    365-402.

10. H. Cohen, "Nonlinear diffusion problems", in Studies in Applied Mathe-
    matics (A. H. Taub, ed.), Englewood Cliffs: Prentice-Hall. 1971,
    27-64.

11. H. Cohen, "Mathematical developments in Hodgkin-Huxley theory and its approximation", in Lectures on Mathematics in the Life Sciences, vol. 8 (S. A. Levin, ed.), Am. Math. Soc., Providence, R.I. 1975.

12. K. S. Cole, Membranes, Ions and Impulses, University of California Press, Berkeley, California. 1968.

13. J. Cooley, F. Dodge, and H. Cohen, "Digital computer solutions for excitable membrane models", J. Cell. Comp. Physiol. 66, Supp 2 (1965) 99-109.

14. J. W. Cooley and F. A. Dodge, "Digital computer solutions for excitation and propagation of the nerve impulse", Biophys. J. 6 (1966) 583-599.

15. J. S. Coombs, J. C. Eccles, and P. Fatt, "The electrical properties of the motoneuron membrane", J. Physiol. (Lond.) 130 (1955) 291-325.

16. F. A. Dodge and J. W. Cooley, "Action potential of the motoneuron", IBM J. Res. Devel. 17 (1973) 219-229.

17. J. C. Eccles, The Physiology of Nerve Cells, The Johns Hopkins Press, Baltimore, Maryland. 1957.

18. J. C. Eccles, The Physiology of Synapses, Academic Press, New York. 1964.

19. J. W. Evans, "Nerve axon equations: I. Linear approximations", Indiana Univ. Math. J. 21 (1972) 877-885.

20. J. W. Evans, "Nerve axon equations: II. Stability at rest", Indiana Univ. Math. J. 22 (1972) 75-90.

21. J. W. Evans, "Nerve axon equations: III. Stability of the nerve impulse", Indiana Univ. Math. J. 22 (1972) 577-593.

22. J. W. Evans, "Nerve axon equations: IV: The stable and the unstable impulse", Indiana Univ. Math. J., to appear.

23. R. FitzHugh, "Impulses and physiological states in models of nerve membrane", Biophys. J. 1 (1961) 445-466.

24. R. FitzHugh, "Mathematical models of excitation and propagation in nerve", in Biological Engineering (H. P. Schwan, ed.), McGraw-Hill, Inc., New York. 1969, 1-85.

25. S. P. Hastings, "The existence of periodic solutions to Nagumo's equation", Quart. J. Math., Oxford 25 (1974) 369-378.

26. A. L. Hodgkin, The Conduction of the Nerve Impulse, Charles C. Thomas, Springfield, Illinois. 1964.

27. A. L. Hodgkin and A. F. Huxley, "A quantitative description of membrane current and its application to conduction and excitation in nerve", J. Physiol. (Lond.) 117 (1952) 500-544.

28. A. L. Hodgkin and W. A. H. Rushton, "The electrical constants of a crustacean nerve fibre", Proc. Roy. Soc. B 133 (1946) 444-479.

29. A. F. Huxley, "Can a nerve propagate a subthreshold disturbance?" J. Physiol. (Lond.) 148 (1959) 80-81P.

30. R. Iansek and S. J. Redman, "The amplitude, time course and charge of unitary excitatory post-synaptic potentials evoked in spinal motoneurone dendrites", J. Physiol. (Lond.) 234 (1973) 665-688.

31. J. J. B. Jack, S. Miller, R. Porter, and S. J. Redman, "The time course of minimal excitatory post-synaptic potentials evoked in spinal moto-neurones by group Ia afferent fibres", J. Physiol. (Lond.) 215 (1971) 353-380.

32. J. J. B. Jack and S. J. Redman, "The propagation of transient potentials in some linear cable structures", J. Physiol. (Lond.) 215 (1971) 283-320.

33. J. J. B. Jack and S. J. Redman, "An electrical description of the moto-neurone, and its application to the analysis of synaptic potentials", J. Physiol. (Lond.) 215 (1971) 321-352.

34. M. Kuno, "Quantum aspects of central and ganglionic synaptic transmission in vertebrates", Physiol. Rev. 51 (1971) 647-678.

35. L. Davis and R. Lorente de Nó, "Contribution to the mathematical theory of the electrotonus", Studies from Rockefeller Inst. for Med. Res. 131 (1947) 442-496.

36. H. D. Lux, P. Schubert, and G. W. Kreutzberg, "Direct matching of morpho-logical and electrophysiological data in cat spinal motoneurons", in

Excitatory Synaptic Mechanisms (P. Anderson and J. K. S. Jansen, eds.)
Universitetsforlaget, Oslo, Norway. 1970, 189-198.

37. H. P. McKean, "Nagumo's equation", Advances in Mathematics, 4 (1970)
     209-223.

38. L. M. Mendell and E. Henneman, "Terminals of single Ia fibers: location,
     density, and distribution within a pool of 300 homonymous motoneurons",
     J. Neurophys. 34 (1971) 171-187.

39. J. S. Nagumo, S. Arimoto, and S. Yoshizawa, "An active pulse transmission
     line simulating nerve axon", Proc. IRE. 50 (1962) 2061-2070.

40. P. G. Nelson and H. D. Lux, "Some electrical measurements of motoneuron
     parameters", Biophys. J. 10 (1970) 55-73.

41. W. Rall, "Branching dendritic trees and motoneuron membrane resistivity",
     Exp. Neurol. 1 (1959) 491-527.

42. W. Rall, "Theory of physiological properties of dendrites", Ann. N. Y.
     Acad. Sci. 96 (1962) 1071-1092.

43. W. Rall, "Theoretical significance of dendritic trees for neuronal
     input-output relations", in Neural Theory and Modeling (R. F. Reiss, ed.)
     Stanford University Press, Stanford, California. 1964, 73-97.

44. W. Rall, R. E. Burke, T. G. Smith, P. G. Nelson, and K. Frank, "Dendritic
     location of synapses and possible mechanisms for the monosynaptic EPSP
     in motoneurons", J. Neurophys. 30 (1967) 1169-1193.

45. W. Rall and G. M. Shepherd, "Theoretical reconstruction of field poten-
     tials and dendrodendritic synaptic interactions in olfactory bulb",
     J. Neurophys. 31 (1968) 884-915.

46. W. Rall, "Time constants and electrotonic length of membrane cylinders
     and neurons", Biophys. J. 9 (1969) 1483-1508.

47. W. Rall and J. Rinzel, "Branch input resistance and steady attenuation
     for input to one branch of a dendritic neuron model", Biophys. J. 13
     (1973) 648-688.

48. S. J. Redman, "The attenuation of passively propagating dendritic

potentials in a motoneurone cable model", J. Physiol. (Lond.) $\underline{234}$ (1973) 637-664.

49. J. Rinzel and W. Rall, "Transient response in a dendritic neuron model for current injected at one branch", Biophys. J. $\underline{14}$ (1974) 759-790.

50. J. Rinzel and J. B. Keller, "Traveling wave solutions of a nerve conduction equation", Biophys. J. $\underline{13}$ (1973) 1313-1337.

51. J. Rinzel, "Spatial stability of traveling wave solutions of a nerve conduction equation", Biophys. J., to appear.

52. J. Rinzel, "Neutrally stable traveling wave solutions of nerve conduction equations", preprint (1975).

53. J. Rinzel, "Voltage transients in neuronal dendritic trees", Federation Proc. $\underline{34}$ (1975) 1350-1356.

54. N. H. Sabah and K. N. Leibovic, "The effect of membrane parameters on the properties of the nerve impulse", Biophys. J. $\underline{12}$ (1972) 1132-1144.

55. G. M. Shepherd, The Synaptic Organization of the Brain, Oxford University Press, New York. 1974.

56. R. B. Stein, "The frequency of nerve action potentials generated by applied currents", Proc. R. Soc. Lond. B $\underline{167}$ (1967) 64-86.

57. C. F. Stevens, Neurophysiology: A Primer, John Wiley, New York. 1966.

58. W. C. Troy, "Oscillation phenomena in nerve conduction equations", Doctoral thesis, State University of New York at Buffalo. 1974.

59. S. Yoshizawa, "Population growth process described by a semilinear parabolic equation", Math. Bios. $\underline{7}$ (1970) 291-303.

Mathematical Research Branch, NIAMDD
Building 31, Room 9A-17
National Institutes of Health
Bethesda, Maryland 20014

Lectures on Mathematics in the Life Sciences
Volume 8, 1976

IS NEUROBIOLOGY TOTALLY NON–MATHEMATICAL?

H B Barlow[1]

ABSTRACT

Mathematics has not contributed very much to the great increase of knowledge of the nervous system over the past twenty years, but this may result from mathematicians not having the necessary background of neurobiological knowledge. Von Neuman pointed to the limited accuracy of nerve cells as an interesting and important limitation; recent advances in this field are reviewed, and an experimental attempt to measure the statistical efficiency of human subjects performing perceptual tasks is described.

1. Necessity of neurobiological knowledge

The title of my talk arose from the following thought. Neuro-biology has advanced strikingly over the past twenty years, and I think it would be possible to get pretty good agreement that these advances have taken place, what they are, who made them, and how. You could then ask the question "How many of these advances involved the use of anything more than High School Mathematics?". I can only think of two or three: the mathematics of nerve excitation, conduction, and membrane equilibria; the statistics of release of transmitter substance in finite packages; and possibly a third is the application of signal detection theory to sensory processes, with which I shall mainly be concerned later on.

Most of the important advances that have taken place involve virtually no mathematics. These have depended on relating physiological function to anatomical structure, or to pharmacology, or to biochemistry –

[1] Address: Physiological Laboratory, Cambridge CB2 3EG, England. This research was supported by Grant EY 00276 from the U S P H S.

an increase of detailed knowledge of how nerve cells are interconnected, how they grow, the substances they secrete and so forth. These involve complex processes that haven't yet been handled quantitatively or mathematically.

I can imagine someone saying "But wasn't the start of all this the solution of the problem of excitation and propagation - the Hodgkin Huxley equations?". There is something in that, possibly, but I could reply that the establishment of chemical transmission at synapses by Loewe, Dale, and others was just as important and involved only Junior High arithmetic. And I think it could be held that the solution of propagation really resulted from a physical insight, not a mathematical one. The solutions came from seeing the nature of the physical problem (cable structure, non-equilibrium situation for Na^+ ions providing a local source of energy, and so on). The mathematics mainly followed from this and I don't think it gave the initial insight. In fact I doubt if mathematical insight has contributed critically to the solution of any biological problem. Having said this, I must add that I very much hope I shall be proved wrong tomorrow, or even today!

Now in other sciences, a mathematical formulation or insight has done this, why not in neurobiology? One reason, it seems to me, is because mathematicians in neurobiology have always been the handmaidens of applied physicists, or worse still, of a neurobiologist. A mathematician can be extraordinarily useful in this position, but no more than useful. To contribute major insight, you must spot the major problem, and to do that you cannot take your guidance from someone else. You cannot trust experts, and simply have to understand the evidence yourself. And that may take two to five years, by which time you will probably have forgotten the original intuition that told you "There are interesting problems over there"! I suppose it is the nervous system's capacity to produce intelligent thought and behaviour that especially intrigues mathematicians. What perhaps might intrigue them equally is the fact that it does this with elements that obviously have severe limitations that ought to be definable. It seems to me an enviable and challenging task for any mathematician to learn neurobiology keeping always in mind the two thoughts "But these assemblies of nerve cells can think intelligently" and "What are the basic limitations to what a nerve cell can do?". And there is a third aspect you probably also have to learn about, namely that the brain develops into its adult state over a considerable time - decades for the human. So a neuromathematician also needs to know the whole of developmental biology.

John Von Neuman was intensely interested in the brain for the last
5 or 10 years of his life, and one of the apparent limitations of nerve
cells that intrigued him was their low metrical accuracy. He judged
that they had only 2 decimal digit accuracy, whereas he thought something
approaching 10 digit accuracy would be desirable in a computer that
performed brain-like tasks (Von Neuman, 1958). I thought it might
interest you if I now review what we know about this. I shall talk about
the reliability of sensory nerve cells, and compare it with the
reliability of overall detection processes in humans. Our views have
undergone something of a change on this topic over the last few years.
It used to be supposed that an individual cell was an unreliable
component, and that reliability was achieved by using many in parallel.
This is probably not so on the sensory side, and on the motor side as
well evidence has been obtained showing how quite finely graded
contractions of a whole muscle can be achieved when each individual
nerve fibre seems only able to contribute a series of more or less
spasmodic or tremulous jerks to the muscle's movement. This knowledge of
input and output does not of course tell one how the brain makes up for
those missing 8 digits of accuracy, but we are beginning to see how
nerve cells are actually used in elementary tasks.

2. Accuracy and reliability of sensory performance

If you measure the intensity of a stimulus that a human subject is
observing, you characteristically find that there is rather a large
range of physical values which the subject sometimes detects and sometimes does not. This range is not a matter of 5 or 10%, but typically
involves a change of the stimulus intensity by a factor 2 or 3, and it
used to be attributed to "biological variability": the notion was that
some internal "threshold level" in the observer fluctuated, so that the
same physical stimulus sometimes exceeded it and was observed, and sometimes did not. This notion was modified abruptly about 30 years ago when
Selig Hecht discovered that the cause of fluctuations in visibility of
weak lights was not biological but physical, and resulted from the quantum
nature of light. This work is probably well known to some of you, but I
shall take it as my starting point, and then discuss experiments on single
ganglion cells which seem to provide a satisfactory physiological basis
for the overall performance of human subjects. Finally I shall discuss
some psychophysical experiments I have started to do which will, I hope,
lead to understanding more complex aspects of perception.

3. Human absolute threshold

Hecht, Shlaer and Pirenne in about 1940 decided to reexamine the
question of the absolute threshold of vision in humans - that is to
find out the minimum quantity of light that gives a sensation of light.
This involved preliminary experiments to find the optimum conditions:
what is the best wavelength to use? Whereabouts on the retina should the
image fall? How large should it be? How long should it last? How often
can you repeat it? And so on. I shall not go into these particulars
because the results conformed with what was generally accepted. When
they made measurements on skilled, attentive subjects under optimum
conditions the result they obtained was a good round number - about 100
quanta at the cornea. They then proceeded to estimate how many of these
quanta were likely to be absorbed in receptors, and came up with another
round answer: under 10. I may as well say at once that, although the
arguments they used were skilful and persuasive until recently, this
answer is probably wrong - by a factor of 2 or 3 - so that 20 or 30
quanta are probably absorbed at threshold. But in spite of this rather
big error much of the interest of the rest of their argument remains, and
I shall present it. Thinking that only 10 were absorbed, and perhaps not
all of them effective, the first conclusion they drew was that there were
not enough quanta for any single receptor cell to absorb more than 1
quantum. The reason is that their optimum stimulus covered several
hundred receptor cells, and it could actually have been spread out to
cover several thousands without greatly increasing the number of quanta
required for a visible flash. So this conclusion stands even though 20
or 30 were actually absorbed from a threshold flash. And this conclusion
conveys the general message of the first part of my talk, namely
"Biological receptors and nerve cells are more sensitive and reliable
than you might expect".

Their next argument was as follows. If only 10 critical events
underlie seeing a flash of light, there is bound to be considerable
variability of response, because there are few events, they are
independent, and each is subject to Poissonian variability. The
probability of an event depends upon the intensity of light in the flash,
and they showed how the probability of 1 or more, 2 or more, etc. varied
with log intensity. But they already knew how the probability of seeing
varied with log intensity, so they could simply see which of these curves
matched their results, and this critical number of events for seeing
appeared to be 6, 7 or 5 in three experiments. Those figures are a bit
below their estimate of "under 10" absorptions at threshold, but it was

close enough for them to be satisfied and to conclude that the major
cause of variable seeing in the threshold range was physical in origin,
not biological. (For more details, and figures, see Hecht, Shlaer and
Pirenne, 1942, or Pirenne, 1967).

Nowadays we have to worry why the curves have slopes characteristic
of responses dependent on 5 to 7 critical events when we believe 20 to 30
quantal absorptions occur. I don't want to go deeply into this, because
I think there are more interesting lines to explore, but I will indicate
the direction thought is going about this problem.

First, one thinks now of "detecting" a flash rather than "seeing" it.
Thresholds are out, and one thinks in terms of signal detection theory.
This leads very directly to the concept of intrinsic noise - the internal
disturbance that makes it difficult to detect sensory stimuli. And there
is another definable characteristic of the detection process: What
proportion of the statistical information available does it actually make
use of? There is little doubt that intrinsic noise and variable
statistical efficiency will enable one to fit the facts of absolute
threshold, but it is another matter whether the measure of intrinsic
noise and statistical efficiency that one emerges with will be useful in
other contexts. I shall return to efficiency estimates later, but first
plan to look at the neurophysiology of detecting weak light flashes.

4. Absolute threshold of cat ganglion cells

What sort of changes occur in the nerves leading to the brain when a
just visible flash of light is seen? One can get some information about
this by recording from the ganglion cells, which send their long axons
centrally to the brain. The receptors in the retina absorb the quanta.
This influences the membrane potential, which changes the amount of
transmitter released at the receptor terminal, and thus changes the
membrane potential of the bipolar cell, and through a similar sequence we
finally reach the ganglion cell. Here something different happens:- the
change has to be transmitted, not for 1/5 mm from receptors through the
retina, but for 50 or 100 mm to the next relay point in the brain, the
Lateral Geniculate nucleus, and it is at the ganglion cell that the all-
or-none action potentials that are transmitted along the optic nerve to
the brain are generated. I've chosen to tell you about the responses
here, rather than at an earlier or later point in the pathway, first
because the action potentials are detectable extra-cellularly in mammals,
so they can be studied under conditions not too different from those of
Hecht's experiments. Second, the train of action potentials has actually

been so studied at this point, whereas comparable studies have not been
carried out at more central points and are only now being carried out
more peripherally. The work I shall talk about was done with Levick and
Yoon on cats, and as previously I shall spend no time on the methods used,
but ask you to accept that what we record is not too different from what
normally goes on. Bear in mind that I am talking of cat, though I hope
it isn't too different from the human.

When you record from a typical ganglion cell there is a maintained
discharge, and this is found even when the cell is in total darkness.
Upon stimulation a brief burst of extra impulses occurs, and if the
stimulus is repeated several times an average response, or Post Stimulus
Time Histogram, can be built up. Repetition is desirable because of the
variability, and with a mini-computer it is relatively easy to get a
measure of the variability as well as the average response.

Figure 1 shows the result of an actual experiment. The retina was
in darkness, and the ganglion cell had been selected from the most
sensitive region. The top record shows the PSTHS to a weak flash. Note
the irregular maintained discharge and the response, which appears rather
big in this average of 100 repetitions. Below are shown the Pulse Number
Distributions - that is the distribution of the numbers of impulses
during this time interval both when a stimulus was, and when it was not,
delivered. Now you see that the distributions overlap grossly, and even
though the response looked large in the PSTH, it could not have been
detected on every repetition. A crude calculation, indicated in Fig. 1,
suggests that the stimulus would have had to be about 3 times stronger
for reliable detection. However only 5 quanta were delivered on average
in this experiment, so the threshold of this ganglion cell, by this
criterion, would be only 15 quanta at the cornea. Ten other ganglion
cells gave figures ranging from 9 to 70 quanta at the cornea. The one I
have chosen to illustrate has a threshold very close to the average of the
five most sensitive units.

Remembering that humans need about 100 quanta at the cornea, this
figure of 15 for the threshold of a cat's ganglion cell is possibly about
what you would expect if you make a couple of assumptions. The first is
trivial - that cats, are a bit more sensitive than humans. But the
second is crucial for the topic I am discussing, namely the accuracy and
reliability of nerve cells. This second assumption, which is supported
by these results, is that threshold responses are mediated by the most
sensitive cell, not by averaging lots of cells.

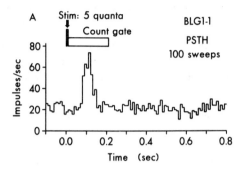

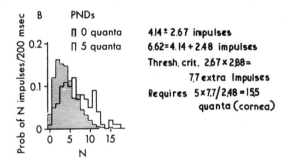

Figure 1A. The average response of a fully dark adapted cat retinal ganglion cell to light delivered to its receptive field. Impulses were collected in 10 msec bins before, during, and after 100 repetitions of the weak, brief, flash, and the numbers are expressed as impulses/sec. Below (B) are the distributions of the numbers of impulses in the 200 msec periods following the stimulus (RH distribution) and preceeding the stimulus (LH distribution). An average of 2.48 impulses resulted from the 5 quanta delivered to the cornea; the criterion, corresponding to S/N ratio of 2.88, would be 7.7 extra impulses, for which 15.5 quanta at the cornea would be required.

If that is right, it is very important. Behaviour does not emerge
as a property of the millions of cells in the brain in the way that
pressure emerges as a property of a gas. I think this is intuitively
obvious to some people, but it isn't to others, and the contrary view was
accepted by many neurophysiologists in Von Neuman's day, which is why I
emphasise it.

Although I have already jumped to the main conclusion from this work
on cat ganglion cells, the analysis can be carried further, and I want to
indicate briefly what was done. More details can be found in Barlow,
Levick and Yoon (1971).

If 15 quanta fall on the cornea of the cat's eye, how many are
absorbed? This, you remember, was where Hecht slipped up. We were a bit
more cautious in our estimates, and concluded that it could not be more
than 7, and was probably about 4. We were therefore, at first, astonished
to see that 4 quantal absorptions caused 8 extra impulses. How could 1
quantum cause 2 impulses? Or was our higher estimate actually the correct
one? This is a question that a statistical analysis of the results might
help to answer, so we proceeded to that.

The mean number of extra impulses varies linearly with average number
of corneal quanta at low intensities, though it starts to saturate if
higher intensities are used. There is a maintained discharge, and knowing
the slope of the relationship between extra quanta and extra impulses, you
can express the maintained discharge as being due to a fictional "dark
light". You can think of this as due to thermal isomerization of photo-
sensitive molecules if you like, though there are several other equally
plausible causes of it.

The variance of the number of impulses in response to different
intensities also varies linearly with intensity, and again there is a
resting variance possibly explicable by "dark light". But the important
result is that variance increases more rapidly than mean response. If
one was dealing with a straightforward random process, such as 1 impulse
for 1 quantal absorption, then the additional variance should be equal to
the amount added to the mean, but it is not. To make use of these results
we derive various quantities. F is the fraction of corneal quanta
utilized in generating impulses. M is the number of impulses per
effective quantum. F and M are calculated from experimental measurements,
and the averages for the 5 best units were $F = 0.13$, $M = 2.8$. We had
other methods of calculating F etc. but I shall not go into them.

Clearly the generation of more than 1 impulse per absorbed quantum is substantiated, and there was an unexpected dividend from this analysis. The distribution of intervals between impulses in the maintained discharge of a cell under steady illumination is quite well fitted by a gamma distribution, and under usual conditions this has a parameter of 2 to 5 (see FitzHugh, 1957; Barlow and Levick, 1969). This makes good sense if one supposes that several preceding events - possibly quantal absorptions - are counted up before the cell discharges, though I should say that there are quantitative difficulties with any such simple hypothesis. In the dark, to our astonishment the impulse interval distribution assumed a very skewed shape; short intervals were commoner even than if it had been exponential, and to our even greater astonishment, this still fitted a gamma distribution, but one with a fractional parameter. This seemed ludicrous at the time, but if we had had more faith in the mathematical implication we might have reached our later conclusion much earlier. Presumably this gamma distribution with fractional parameter is generated by a Poisson process, followed by a mechanism which generates several impulses for each Poisson event. As a crude check on this, we took the original distribution and plotted the intervals between every third impulse. To our delight this produced the exponential distribution expected for the generating Poissonian process. There are snags in the physical interpretation, but I mention this because it is a case where the mathematics suggested insight that we failed to appreciate, and only reached by another route at a much later time.

Now where has all this lead us in relation to Von Neuman's problem about the limited metrical accuracy of nerve cells? I think it is quite interesting. First, he was obviously right in saying that a train of nerve impulses, such as those I have shown you, cannot transmit a quantity with high accuracy. But it is equally clear that this is not actually a limiting factor in the type of situation that we have been considering. The nerve cells are used in such a way that their own inaccuracies are of trivial importance, and what limits performance is variability of the input - in this case quantum fluctuations. That may not always be the case but perhaps the story I have told confirms the notion that nerve cells are so connected up that each one performs a significant task. For many calculations it is convenient to have 10 digit accuracy, but you can do it with 1 or 2 if you perform it a different way. Obviously the main trick here is to ensure that your variables not only remain within their limiting values, but also range over the whole of it. For instance if a variable has a large constant part and small variable quantity added to it, then you should remove that constant if you possibly can before doing your manipulations.

5. Absolute measures of visual efficiency

Just to review: I started by describing the remarkable ability
human subjects have in being able to detect very weak lights; I then
showed that this was based upon the ability of the ganglion cells of the
retina to respond reliably to the numbers of quanta absorbed; and I have
suggested that it is absolutely necessary for nerve cells to respond
directly to the variables we are sensitive to, because of the fact that
Von Neuman picked on, namely their limited metrical accuracy. The n.s.
cannot afford to use their restricted dynamic range for anything but the
relevant parts of the variables.

Can one put this argument into reverse? Can one say "Human subjects
do such and such very well: therefore they must have nerve cells directly
connected to respond to the quantities that are important in doing such
and such"? One can regard colour discrimination as an example. Humans
detect very small wavelength changes: therefore there are some nerve cells
that respond to wavelength change (rather than luminance change as many
others do). As DeValois et al (1967) have shown, this is so.

I want now to describe an attempt, which is still in a preliminary
stage, to find other perceptual tasks which humans "Do well", in order,
hopefully, to get some hints about higher perceptual nervous mechanisms.
Clearly the first problem is to choose tasks that are not obviously
limited by properties of the sense organ or receptors, since we want to
explore more central mechanisms. For this reason I chose the task of
detecting figures in random dot noise. The subject looks at a screen
that has rather the appearance of a t.v. set tuned between stations, and
his task is to detect some image that may emerge from this noise. There
is no question of the visibility of the dots, but the detectability of
the image is another matter.

The second problem is to decide if he is "doing well" at this task.
We need a measure that enables one to compare performance at different
tasks, and if possible we would like to be able to compare the
performance of a single neurone in an animal's brain with that of a
human subject, as we were able to do with absolute sensitivity and
wavelength discrimination. For this I used the measure of statistical
efficiency: "What proportion of the statistical information available to
the subject did he actually use?" I shall show how this is done, but the
actual answers I have to report at this stage are somewhat meagre.

The subject looks at a display of a continuously changing array of dots. Adjustment of a comparator to one extreme turns the screen black, except for an occasional bright speck appearing. At the other extreme, the screen appears filled with a regular array of white dots except for the occasional black speck where one is missing. One parameter, the dot probability, specifies the background condition, but of course the perceptual experience of this constantly changing pattern of dots is quite complex.

On pressing a button, the dot probability within a selected region is changed for a selected time, and the subject has to report whether he saw anything. By suitable gating, the actual number of dots is counted and recorded, as well as the subject's responses. The shape and size of the target area is controlled by a flying spot scanner. Thus the subject is performing a task rather like a normal visual threshold, except that the dots which he sees are all above threshold, and in principle separable from each other in space and time. One hopes that the limit to performing such a task lies in the central analysing mechanisms rather than peripherally.

Now let me show the result of a typical experiment. In Fig. 2 the target was a small dot, appearing for a variable duration. The subject adjusted the dot density in the target area, and what is plotted here is the log of the average added number of dots required as a function of log duration. Also plotted are two properties of the number of dots in the target area, over a time equal to signal duration, since these numbers characterise the background against which the subject must pick out the target. Top is N, the average number, and bottom is σ_N, the standard deviation of the distribution of these numbers. Because the distribution is binomial this has a value slightly below $\sqrt{N}$, and it rises, on this log.log. plot, with a slope of $\frac{1}{2}$.

You might suppose either N or σ_N was the factor limiting ΔN; one is so used to sensory thresholds rising with the magnitude of the stimulus that it is natural to expect the kind of rise of ΔN shown. On the other hand σ_N really imposes a limit on what it is possible to detect; no observer could reliably detect an increment that lay within the distribution of increments that occur frequently by chance, for if he did so he would give very frequent false positive responses. The next figure throws light on whether N or σ_N is the important factor.

In Fig. 3 the target duration and area were constant, but the background dot probability was varied. Here the scales are linear, and

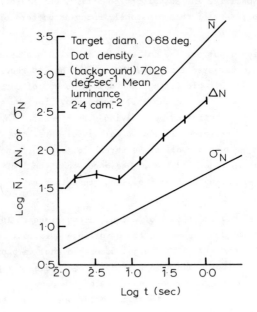

Figure 2. The effect of exposure duration (t) on the average number of extra dots (ΔN) required for detection against a fluctuating background of dots. Lines labelled $\overline{N}$ and σ_N show the average number of dots in the background within the target area and duration, and the standard deviation of this number.

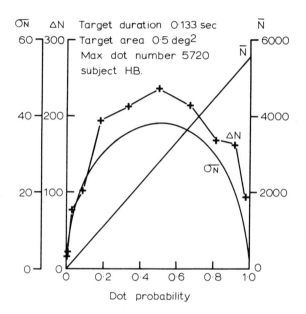

Figure 3. The effect of varying the average number of dots in the background on the number of added dots required for detection (ΔN). N is the average number of dots (scale on right) and is a linear function of dot probability (abscissa). Because the distribution in the background is binomial, σ_N is an inverted U, and ΔN is a fairly constant multiple of σ_N.

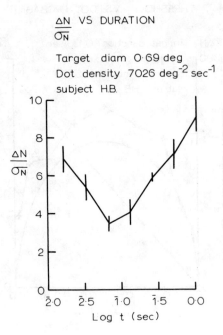

Figure 4. The effect of exposure duration on $\Delta N/\sigma_N$, the signal/ noise ratio at threshold in the external stimulus (extrinsic S/N). Results of Fig. 2 replotted.

you see the inverted U of the experimental results for the observer's
threshold. N of course rises linearly with dot probability, but σ_N,
for the binomially distributed dot numbers, also follows an inverted U.
Note that the scales for ΔN and σ_N are different: ΔN is always 5 or 6
times σ_N.

In Fig. 4 the first results are replotted as $\Delta N/\sigma_N$, the apparent,
or extrinsic, S/N ratio required at threshold. Notice that there is a
fairly sharp minimum at about 1/10 sec. If the stimulus is shorter or
longer you require a larger S/N ratio. If area instead of time is changed,
one again finds that the extrinsic S/N ratio changes. For instance with
a 130 msec stimulus, other conditions as in Fig. 4, extrinsic S/N ratio
of 5 was needed for a 24 min. spot, but it rose to 18 for a 3° spot.

So now we know that in performing such a task a subject requires an
added number of dots that is several times the noise level. This S/N
ratio varies with target size and duration, and actually it also varies
with the total number of dots. What we cannot tell from the results I
have so far shown is whether a subject sometimes needs a very high
extrinsic S/N ratio because he is insisting on very high reliability, or
whether he requires a high signal strength simply because he is utilizing
the available information very inefficiently. Are the distributions of
the neural representation of the dot images as narrow as they are
extrinsically, or have they been broadened by the sensory processes so
that a bigger signal strength is required to separate them?

How can one distinguish what is happening? Counting false positives
is a possibility, but on either hypothesis we expect them to be < 1%, so
it would be tedious to count them, and a high rate might anyway be
explained by the occasional lapses of attention that we all know occur.
So the technique I used was as follows. Present stimuli of various
intensities, and determined the proportion of successful detections -
do a frequency-of-seeing curve in fact. This should correspond to the
integral of either the narrow extrinsic distribution or the broadened
internal representation, and we should easily tell which it is.

It sounds easy, but the analysis was troublesome until I found a
trick, which I am sure some of you are already aware of. All these
binomials have different slopes, so each experiment was a special case,
but a simple transformation of the abscissa makes them into a family of
nearly parallel curves. The transformation is $Z = n^{\frac{1}{2}}\sin^{-1}(2P - 1)$, and
this has the effect of making the maximum slope of cumulative binomials
equal to $1/\sqrt{2\pi}$.

Now if one plots the experimental curves on these transformed coordinates, if the subject is performing ideally, his frequency of seeing curves will have a maximum slope of $\sqrt{2}$ like that of a true binomial distribution. But if he has degraded those narrow distributions by counting only a fraction of the dots, or broadened them by any other means, then his frequency of seeing curves will be flattened. In fact the flattening gives a very direct measure of the degradation, and it is convenient to represent this as a loss of statistical efficiency: it is "as if" he had only used a certain fraction, F, of the information available.

Now you remember I drew your attention to two conditions, one where the S/N ratio was about 5, the other where it was about 18. Figure 5 shows frequency-of-seeing curves determined under the two conditions, and plotted on these transformed coordinates. Both are shallower than the ideal curves would be, but the one where the S/N ratio was 18 is much flatter than the one where it was 5. The appropriate values of F are indicated here.

This suggests that what causes variations of $\Delta N/\sigma_N$, which I called the extrinsic S/N ratio, is variation of efficiency; in that sense it is misleading to call it variation of S/N ratio - it is variation of apparent S/N ratio, of the S/N ratio required externally in the stimulus. But the beneficial effects of high S/N ratio are <u>not</u> utilized; the information is wasted, and the <u>internal</u> S/N ratio is much lower.

I think the technique is paying off. Sometimes the perceptual system is efficient, but under other conditions it is not, and I can't help hoping that what we perceive efficiently will reveal something about the neural mechanism. Because the nerve cells only have Von Neuman's 2 digit accuracy, doing it well probably means doing it directly - that is having nerve cells that are themselves selectively sensitive to the stimulus that is efficiently detected.

6. Conclusions

Whether or not this approach gives us interesting quantitative information about perceptual abilities such as detecting motion, or symmetry, or other patterns of stimulation, there is one result that is relevant to other material I have discussed. You remember I started from Hecht's experiments on absolute threshold, but had to admit that he made a mistake in supposing that quantum fluctuations accounted for most of the variations in visibility of weak flashes, because he underestimated

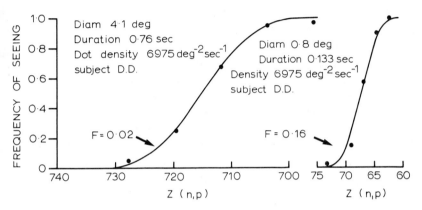

FREQUENCY OF SEEING CURVES

Figure 5. Frequency of detection as a function of the Z-transform of
the number of dots added in the stimulus: $Z(n, p) = n \cdot \sin^{-1}(2p - 1)$.
For 100% efficient detection the maximum slope is always $1/\sqrt{2\pi}$, and for
reduced efficiency it is diminished by $\sqrt{efficiency}$.

the fraction of quanta absorbed. Now so far I have never found any task
that subjects can perform with efficiency greater than about $\frac{1}{2}$, and I
think this is likely to be the factor that reconciles Hecht's results
with the latest estimates of quantal absorption. I expect this loss of
efficiency occurs at the decision-making stage, because the optic nerve
fibres seem to be conveying good, un-degraded, information to the brain,
and this is indeed one example showing how nerve cells can behave
physically as more reliable elements than was popularly supposed in Von
Neuman's day.

Finally let me retrace my steps right back to the beginning to
suggest that the reason neurobiology appears so unamenable to mathematics
is simply because mathematicians have not found out enough about neurobiology. The available information must be filtered through
mathematical brains, for non-mathematical neurobiologists are likely to
filter-out the insight you might have if you received all the information.

Neurobiology is <u>almost</u> totally non-mathematical simply because
mathematicians are <u>almost</u> totally non-neurobiological.

REFERENCES

Barlow, H.B. and Levick, W.R. (1969). Changes in the maintained discharge
  with adaptation level in the cat retina. J. Physiol. (Lond) 202, 699-
  718.
Barlow, H.B., Levick, W.R. and Yoon, M. (1971). Responses to single
  quanta of light in retinal ganglion cells of the cat. Vision Res.
  (Supplement No. 3) pp 87-101. Pergamon Press 1971.
DeValois, R.L., Abramov, I. and Mead, W.R. (1967). Single cell analysis
  of wavelength discrimination at the lateral geniculate nucleus in the
  Macaque. J. Neurophysiol. 30, 415-433.
FitzHugh, R. (1957). The statistical detection of threshold signals in
  the brain. J. Gen. Physiol. 40, 925-948.
Hecht, S., Shlaer, S. and Pirenne, M.H. (1942). Energy, quanta, and
  vision. J. Gen. Physiol. 25, 819-840.
Pirenne, M.H. (1967). Vision and the Eye. Science Paperbacks.
Von Neuman, J. (1958). The Computer and the Brain. Yale Univ. Press,
  New Haven.